THE CLAIM OF CROFTING

THE CLAIM OF CROFTING

THE SCOTTISH HIGHLANDS AND ISLANDS, 1930-1990

JAMES HUNTER

MAINSTREAM
PUBLISHING
EDINBURGH AND LONDON

First published in Great Britain 1991 by
MAINSTREAM PUBLISHING COMPANY (EDINBURGH) LTD
7 Albany Street
Edinburgh EH1 3UG

ISBN 1 85158 329 7 (cloth)

A catalogue record for this book is available from the British Library

The publisher gratefully acknowledges the financial assistance of the
Scottish Arts Council in the publication of this volume

Typeset in 11/13 Goudy by Blackpool Typesetting Services Ltd, Blackpool
Printed in Great Britain by Biddles Ltd, Guildford

For
Angus Macleod
but for whom there would have been no
Scottish Crofters Union

CONTENTS

ACKNOWLEDGMENTS

Many people and many organisations have helped make this book possible. The Scottish Crofters Union was kind enough to commission it. The Highlands and Islands Development Board and the Royal Bank of Scotland generously assisted the SCU to meet the cost of that commission. And I trust that the royalties which the union will receive on sales will make the venture worthwhile from the SCU's point of view.

Among the many institutions whose staffs dealt courteously and efficiently with my requests for archival material were the Scottish Record Office, the National Library of Scotland, Inverness Public Library, Aberdeen University Library, Sabhal Mor Ostaig and the Museums Service of Skye and Lochalsh District Council.

John Bryden made a number of extremely useful comments on those parts of the book dealing with HIDB policy. John Bannerman and Margot MacGregor supplied material that would not otherwise have been available to me.

My wife Evelyn and our children, Iain and Anna, once more gave me the support without which it would have been difficult to pursue my enthusiasm for crofting matters.

Since this book would not have been written but for my association with the SCU, it is appropriate here to express my gratitude to the countless crofters who have so hospitably received me, and so illuminatingly discussed crofting issues with me, in so many parts of the Highlands and Islands in the course of the last six or seven years.

From all the many crofting homes where I have been warmly welcomed, I wish to single out one – for reasons which will be understood by everyone involved in setting up the Scottish Crofters Union. Angus and Annie Macleod have generously fed, entertained and otherwise provided for me during all my frequent visits to Lewis since 1984.

And Angus, with whom I have talked about crofting into the early hours of more mornings than was probably good for either of us, has been an unfailing source of information, good sense and sound advice. To him this book is most respectfully and affectionately dedicated.

INTRODUCTION

One morning in the spring of 1984, at my home in Aberdeenshire, I received a telephone call from John Angus MacKay, then the Highlands and Islands Development Board's representative in Stornoway. The HIDB, John Angus told me, was considering a proposal from the Lewis-based Federation of Crofters Unions. The Federation, a voluntary organisation with no paid staff and almost no financial resources, was of the opinion that the Crofters Union movement should be placed on a new, more professional footing. And the Federation's chairman, Angus Macleod, had consequently asked the Highland Board to help fund the appointment of a full-time union organiser.

There had been a fairly cautious response to this suggestion, John Angus explained. But the HIDB had been sufficiently impressed by Angus Macleod's arguments as to be thinking of putting up the cash needed to commission a modestly-priced feasibility study of the projected reform. Would I, John Angus asked, be interested in conducting such an exercise?

I said I would. And in so saying, though I did not appreciate the fact for some time afterwards, I set my life, and the lives of my family, on a wholly new course.

Since I was then earning the greater part of my living from freelance journalism, I was in a position readily to set aside the month or so that would be required to deal with the task which John Angus MacKay had outlined. But I possessed few other very obvious qualifications for the job.

I was not, and had never been, a crofter. I was neither an economist nor an agriculturalist. And though I had written fairly extensively about Highlands and Islands matters, particularly in the course of a five-year stint on the staff of the *Press and Journal*, the north of Scotland's principal daily newspaper, I had been extremely critical, from time to

11

time, of the policies pursued by the HIDB. I was consequently regarded by some of that agency's senior staff with a certain mild – or maybe not so mild – suspicion.

For a year or two prior to 1984, on the other hand, I had been acquiring a rather different set of insights into developmental possibilities and priorities as a result of having been engaged as a consultant to Rural Forum Scotland, an alliance of organisations representing people living in the countryside.

My work with Rural Forum, on whose behalf I had investigated a number of agricultural issues, possibly gave me more credibility than I would otherwise have had as an HIDB-financed inquirer into prospects for a revamped Federation of Crofters Unions. And it may also have been of some significance in this connection that I was the author of a fairly detailed historical account of the emergence of the modern crofting system.

Indeed it was because of my historical researches – the results of which were published as *The Making of the Crofting Community* in 1976 – that I had first come to the attention of Angus Macleod of the Federation of Crofters Unions. The Crofters Commission, the official agency in charge of crofting administration, had invited me to talk about the crofting past at one of their regular conferences. At lunch that day I happened to sit beside Angus. I thus got to know the Federation chairman just a little bit. And I was soon to get to know this very energetic Lewisman a good deal better; for, having once accepted the HIDB's invitation to report on the practicability of Angus Macleod's reform plans, I found myself drawn more and more into the business of making those plans a reality.

I began – as practically everyone else had done – by being more than slightly sceptical about Angus Macleod's proposals. His notion of equipping the Federation of Crofters Unions with a full-time organiser obviously made a lot of sense. Such an appointment was clearly essential if the Federation was to be made capable of putting the crofting point of view authoritatively and effectively to Government. Nor was it by any means impossible, I quickly gathered, that the Highland Board would agree to provide the Federation with some proportion of the money required to employ a salaried official.

But any such assistance was most unlikely to exceed 50 per cent of the total annual expenditure needed to finance such a venture. The remaining costs would have to be borne by the Federation. Since HIDB aid of this type was always strictly limited in duration, the Federation of Crofters Unions, which had previously had difficulty in scraping a few pounds together, would very soon become solely liable for the substantial and continuing outlays that would be incurred if, as its chairman clearly intended, the Federation was rapidly to transform itself into a professionally managed and genuinely representative organisation of a type which crofters had never previously possessed.

Would several thousand crofters – comparatively few of whom had shown any very great willingness to assist financially with the maintenance of the various local groups which together constituted the Federation of Crofters Unions – really be prepared to make the much larger annual payments required to fund the new departure which Angus Macleod was urging on them? Frankly, I doubted it.

But a three-week tour of the crofting areas in September 1984 persuaded me that I was wrong. Among crofters themselves, I became increasingly convinced, there was no lack of potential support for the Angus Macleod approach. In my report to the Federation of Crofters Unions and the Highland Board I duly recommended that the Federation should indeed reconstitute itself as a single, centrally-directed concern – with its own paid staff and with a membership drawn, ideally, from right across the Highlands and Islands.

This recommendation was accepted at a Federation meeting in Inverness in November 1984. There was then established a Crofters Union Steering Group – with Angus Macleod as chairman and with myself as part-time secretary and co-ordinator – whose task it was to ensure that the Scottish Crofters Union, as the new organisation eventually became known, was put in place by the end of 1985.

The work of the group went speedily and well. Union branches were very soon being formed. An SCU constitution was drawn up. Headquarters premises were obtained in Skye. Membership subscriptions began to be collected. HIDB cash-aid was secured for the vital start-up period. By the summer of 1985 the Working Group felt able to advertise the post of SCU director.

I had been asked much earlier if I was interested in the position. I had replied that I was not. Though I had grown up in the West Highlands, I had lived elsewhere for many years. I was not anxious to move permanently to Skye. I did not want to give up the modestly successful writing and broadcasting career I had carved out for myself. I enjoyed self-employment. I did not, in any case, believe that I was the right man for a job which, I thought, should go to someone much more knowledgeable than myself about the day-to-day practicalities of crofting.

But in August 1985, when the Crofters Union Steering Group sub-committee charged with the appointment of the SCU's first director met in Stornoway to consider its next move, I was again urged to apply. This time I said I would at least think about it. I then went off to spend a weekend in Uist where I was due to address still more of the scores of public meetings which the Steering Group had organised with a view to making its aims better known to crofters.

That Sunday, after he had got home from Mass, I was taken on a tour of South Uist by Roddy Steele, a local crofter. Roddy – whose tragically premature death in 1987 was to deprive the crofting community of one of its ablest spokesmen – was the person who, the year before, had most forcefully put to me the case for the creation of a new and stronger union. I naturally took the opportunity to sound him out about my own situation.

'Look,' I said to Roddy, indicating the sheep and cattle browsing in the afternoon sunshine on the Boisdale common grazing, 'I might know the difference between a cow and a bull. But I'd be hard pressed to tell a cast ewe from a gimmer.'

'That,' replied Roddy, who, as I later came to appreciate, scarcely gave a damn for sheep, being a man who much preferred to work with cattle, 'does not matter in the least.'

We went on for a while in silence. 'Roddy,' I asked at last, for it was this, in truth, that worried me no small amount, 'do you think this union of ours is really going to work?'

He took that question much more seriously. 'It will work if crofters make it work,' he said.

And crofters did just that. In April 1990, when I gave up the union directorship, which I had held then for the best part of five years, more than 4,300 crofters were paid-up members of the SCU. The organisa-tion, thanks to them, had been made wholly self-financing. And the Scottish Crofters Union, if not exactly a power in the land, was certainly a force with which the authorities had to reckon.

This autobiographical beginning is by way of declaring my personal interest in what follows. My original book about crofting dealt, for the most part, with events that took place between the 1790s and the 1920s; with the evictions and enforced emigrations which were so characteristic of the first part of that period in the north of Scotland; with the terrible famine of the 1840s; with the emergence, in the 1880s, of the Highland Land League and with the Land League's ultimately successful campaign for security of tenure; with the land settlement programme which, in the opening decades of the twentieth century, went some way to reversing the clearances of 100 or more years before. These were tumultuous happenings. They aroused – and still arouse – strong emotions. It is difficult for a Highlander to write dispassionately about them. And I make no claim to having done so.

But for all that I felt strongly about these various occurrences, I had no direct involvement in them. It is not so with regard to more recent crofting history.

For much of the 1980s, as I have already mentioned, I was employed by the Scottish Crofters Union. And while, in most of what follows, I have tried to keep my own role – such as it was – very firmly in the background, it would be silly to pretend that the ensuing account of what has been going on in crofting areas in the course of the last 60 years has not, to some extent, been coloured by my SCU connection.

Nor is that the only distinction between the present book and its predecessor. When researching nineteenth-century events, I was dealing with episodes which, for all the controversy surrounding many aspects of them, were sufficiently remote in time as to guarantee that the surviving records of the relevant Government agencies and departments could be readily inspected. But in this respect at least, I have now discovered, the aspiring historian of the later twentieth century is a lot less fortunately placed.

British Government archives are normally closed for 30 years. The records of the Crofters Commission, however, are closed for half a century. This is because those records deal occasionally with individual men and women – in a way which Cabinet papers, for example, generally do not.

One cannot, of course, complain about this zealous safeguarding of people's privacy. But it means, curiously enough, that though historians of United Kingdom foreign policy have for several years had unrestricted access to the British Cabinet's thoughts on the epoch-making Suez crisis of 1956, historians of crofting will have to wait until well into the twenty-first century to gain similar insights into what it was that members of the Crofters Commission – established in 1955 – were doing at the same time.

Of the eight following chapters, therefore, only the first three – covering the 1930s, the 1940s and the early 1950s – are in any way illuminated by that detailed knowledge which can only be obtained as a result of being able to inspect the mass of paper which policy formulation, in this country anyway, invariably generates. Subsequent chapters – dealing with the last 35 or so years – depend very largely on the published reports and statements of the several official bodies which have been primarily concerned with Highlands and Islands affairs. And while such material is certainly plentiful enough, experience suggests that its nature is probably such as to ensure that it does not always reveal exactly what the responsible politicians and administrators were about.

Thus it would not be easy to tell from the little that was published at the time that the Conservative and Unionist Ministers in charge of Scottish affairs in the 1930s were seriously contemplating, even then, the creation of a special agency to deal with the north of Scotland's economic problems. And while one might deduce from various policy initiatives launched in the late 1940s by the Scottish Office's Department of Agriculture that its staff were not well disposed to crofting, it is only by scanning the contents of the Department's internal files, now reposing in the Scottish Record Office, that one can begin to discern something of the sheer dislike which senior civil servants clearly felt for a smallholding system so manifestly at odds with the highly productive, capital-intensive agricultural industry they were endeavouring to promote in post-war Scotland.

But why this concentration on the views of Government and its officials in a book ostensibly concerned with crofting? What of crofters themselves, their feelings, their opinions? Are they not of more importance than the policies pursued by Edinburgh-based civil servants to whom a croft was not so much a warmly regarded homestead, more a little piece of land which – in a famously cynical but not imperceptive phrase – was surrounded on all sides by regulations?

Of course crofters matter. But it is unfortunately the case that, for much of the period examined in this book, they did not shape their own history to anything like the same degree as they did, for instance, in the 1880s.

Then crofters were well organised; constantly taking action on their own behalf; confidently setting their own political agenda; obliging Government, if not quite to do their bidding, then certainly to go a long way to meeting their demands.

The principal means by which the crofting community of the 1880s so effectively articulated its collective point of view was the Highland Land League. But for most of the last 50 or 60 years no such mechanism has been available to crofters.

Briefly in the 1930s and rather more enduringly in the early 1960s, Crofters Unions were formed in a conscious effort to fill this particular gap. The SCU, the direct descendant of these earlier unions, and arguably the remoter offspring of the Land League itself, constitutes one more attempt by crofters to exercise a worthwhile influence on the pattern of events.

The reasons for the crofting community's relative quiescence since the two or three years immediately following the First World War – when a spectacular series of illegal land seizures by crofters propelled the Lloyd George Coalition Government into embarking on the exten-sive redistribution of land long occupied by sheep farmers – are explored at some length in ensuing pages. They derive, in essence, from the demoralising and debilitating impact of protracted economic stagna-tion and, above all, from the almost constant depopulation which resulted from the failure of the Highlands and Islands economy, until very recently, to generate employment and income at rates which even began to approach those typically prevailing in the rest of Britain.

It is not at all surprising, therefore, that the emergence of the Scottish Crofters Union, along with the appearance of many other pointers to renewed social and cultural vitality in crofting areas, should have coincided with the economic revival which took place in many parts of the Highlands and Islands during the 1970s and 1980s – one strong indication of this general change in fortunes being the first marked upturn in the region's population to have occurred since the early nineteenth century.

But if the SCU's formation can thus be presented as one more piece of testimony to the effect that matters have at last taken something of a turn for the better in the north of Scotland, that is not to imply that, prior to the union being set up, the crofting community was, in some sense, at risk from forces which, had the SCU not galloped boldly to the rescue, would have necessarily jeopardised its entire existence.

Such was not at all the case. Crofting had been thus endangered in the era of the clearances. But the victories won by the Land League – especially the gaining of security of tenure in 1886 – had ensured the crofting community's survival by making crofters the beneficiaries of a form of legal protection much more comprehensive and irrevocable than was ever made available to any other class of agricultural occupiers in Great Britain.

Crofters knew this very well. That was why the one thing calculated to rouse them to political action reminiscent of that of the Land League years was a threat – from any quarter whatsoever – to the security of tenure principle. And that, in turn, was why the most serious such threat of modern times – the Crofters Commission's suggestion in 1960 that security of tenure should be made less absolute with a view to facilitating the wholesale amalgamation of crofts – precipitated the uproar culminating in the establishment of the Federation of Crofters Unions.

But if crofting endured primarily because of the Crofters Act of 1886, it did so also because there was no widespread desire – except, from time to time, ironically enough, among the people controlling the very institutions, notably the Crofters Commission and the Department of Agriculture, which one might have expected to be most pro-crofter – to do anything calculated to erode the crofting community's unique legal status. Crofters, on the contrary, were very favourably – if somewhat sentimentally and romantically – regarded by the Scottish population at large. And for all the age-old jokes and jibes about the crofting community's supposedly endemic idleness – the defining such tale being the one about the crofter remarking to the Latin American tourist that there is no Gaelic word conveying anything like the sense of urgency implied by the Spanish *manana* – public goodwill towards crofters continues generally to be the rule.

Why this should be so is a complex subject with ramifications extending well beyond this book's comparatively confined scope. It relates, maybe, to the extent – readily confirmed by a glance at the Glasgow or Edinburgh telephone directory – to which even Scotland's city dwellers are descended from Highlanders and Hebrideans. It has something to do, perhaps, with the standard symbols of Scottish national identity – kilts, tartan and the like – being almost all of Highlands and Islands origin. And it may also be due, in part at least, to nothing more complicated than a widespread feeling that crofters have had an extraordinarily raw deal from history.

But whatever its causes, this popular sympathy for crofters is a fact. And it is a fact with evident political results and implications.

'The Highlander has been the man on Scotland's conscience,' said a Labour Scottish Secretary in 1965 when introducing a piece of legislation, the Highlands and Islands Development Act, designed to do a great deal for those rural north of Scotland constituencies which, in the previous year's general election, had mostly voted Liberal or Conservative – and to do nothing at all for those urban centres where Labour candidates had been overwhelmingly returned.

Nor are such altruistic gestures the prerogative of Labour Ministers. In the late 1980s and early 1990s, an era otherwise distinguished by public spending stringency of considerable severity, Tory Secretaries of State for Scotland very willingly provided £9.5 million annually in order to add to the number of Gaelic television programmes. And whatever benefits Malcolm Rifkind and Ian Lang expected to derive from this quite unprecedented generosity to a minority culture, limited largely to a part of the country then wholly bereft of Conservative MPs, party political advantage was surely not very prominent among them.

Here, then, is one possible interpretation of this book's title. The claim of crofting can be understood to refer to that slightly mysterious ethical and political imperative which results in the wider Scottish and United Kingdom population feeling itself obliged to help sustain a few thousand Highlands and Islands smallholding families – their total numbers equivalent to the inhabitants of one of the smaller English county towns – by permitting the crofting localities to have both their own land laws and their own peculiar entitlement to draw on public funds.

Now it can be argued – and indeed it will be so argued later in this book – that crofting is much less heavily subsidised than is often suggested. But that, in the present context, is entirely by the way. Crofters are widely *believed* to be in receipt of quite exceptionally generous subventions from the public purse. Taxpayers, however, have threatened no revolt in consequence. And though a Conservative Government Minister once told an SCU conference that 'crofters

19

cannot expect their privileges to be indefinitely surrounded by a ring fence', even the cost-cutting Tory administrations of the 1980s shied away from any substantial rejection of the longstanding notion that crofting areas merit special treatment.[1]

But for all that it has contributed significantly to the process whereby public funds have been generously applied to many worthwhile north of Scotland purposes, the widespread acceptance of the claim of crofting – in the sense in which that term can be construed to imply some sort of general duty to be nice to crofters – has not been a wholly unmixed blessing.

If the depopulation of crofting localities, as this book maintains, drained those communities of both hope and vigour; if the deliberate neglect and devaluing of the cultural and linguistic heritage of crofters, as this book also maintains, had the effect of depriving those same crofters of both pride and self-esteem; so the ceaseless talk of subsidy helped engender something akin to what, in the period of Margaret Thatcher's political supremacy, was dubbed a dependency mentality. One does not have to be a Thatcherite to think that a bad thing.

For it to be constantly suggested by those in authority that crofting existed only on some sort of grace-and-favour basis – rather than on account of its own intrinsic worth – was infinitely depressing and dis-piriting. And of all the many tasks which the SCU set itself in the course of its first five years, none was more important than the con-certed attempt made by the union to advance a case for crofting that owed nothing to the previously all too prevalent concept of the crofter as a latter-day representative of that class of people known to the Victorians as the deserving poor.

'The SCU is not in the business of extending a begging bowl to Government on behalf of its members,' the union's first president, Frank Rennie, told one of the organisation's early annual conferences. Rather the union aspired to help crofters 'to expand the economy' of the Highlands and Islands by means of 'promoting the enterprise and initiative' which were indubitably to be found in crofting areas – and which had been exemplified, the SCU president continued, in the 'energetic and enthusiastic response of crofters' to initiatives such as the European Community's Integrated Development Programme in the Western Isles.[2]

State agencies and Government, Frank Rennie acknowledged on behalf of the SCU, had an important part to play in crofting affairs:

> But we do not see the Government's role in these matters as being confined to the provision of public funds for crofting purposes. Such funds are certainly important. Equally important, however, is the Government's overall attitude to crofting. And we feel this attitude is not as supportive as it might be.
>
> It is not enough to blame the present Government for that. Senior civil servants, administrators, agricultural economists, opinion-formers of all kinds have long regarded crofting as a decidedly old-fashioned, even out-moded, form of landholding. The part-time croft, in particular, has been seen as having little or nothing to offer in developmental terms. There has been a tendency to promote the amalgamation of holdings. And policies of support for crofting have been justified largely in terms of the special social needs of the remoter parts of the Highlands and Islands.
>
> This has been unfortunate in that crofting support policies have been presented very negatively. The crofter, it has been implied, is a poor soul whose very survival depends on public subsidies. The crofter's ability to help himself, or herself, has consequently seemed extremely limited, even non-existent. Such perceptions have not helped the wider world's image of the crofter. And they have inflicted considerable damage on morale in the crofting areas themselves.
>
> In the SCU's view, however, there need be no special pleading for croft-ing. That approach was possibly unavoidable when national policy for the countryside was concerned primarily with maximising food production. From that standpoint, the case for crofting no doubt seemed rather weak. But now that there is an urgent, and universally recognised, requirement to devise rural policies designed to meet the circumstances created by agricultural overproduction, the case for crofting is an extremely strong one. Far from being a mildly embarrassing relic from the distant past crofting points the way to the diversified rural economy which is being sought on all sides.

These comments constitute another claim of crofting; a claim which the SCU has since gone on to elaborate and expand; a claim that croft-ing has a very great deal to offer in the context of the rural readjustments now being made in every part of Britain; a claim that crofting, by enabling families to combine the management of a piece of land with some other source of income, is a first-rate means of maintaining a worthwhile population in localities which would otherwise be largely deserted; a claim that crofting, a very unintensive form of agriculture, contributes enormously to the safeguarding of some of Europe's most valuable natural environments; a claim that crofting is equally inti-mately linked with the survival of Scotland's ancient and distinctive Gaelic culture; a claim that crofting works; a claim that crofting makes sense; a claim that crofting is something of which crofters everywhere can and should be proud.

That is the claim of crofting to which the title of this book refers. And though it is a claim which, as ensuing chapters all too clearly demonstrate, would not have received much in the way of widespread approbation in the course of most of the last 60 years, it is a claim which will attract, I hope, a lot more recognition in the future.

Chapter One

BETTER TIMES, WORSE TIMES

In the Scottish Highlands and Islands, as in much of the rest of Britain in this the sixth winter of the country's second war with Germany, 1945 began cheerlessly. 'New Year's Day in Skye,' recorded island diarist Donald Gillies, 'was ushered in with less local celebration than in former years.' Victory, at last, seemed near. Morale was high. But there was, unfortunately, a marked 'scarcity of things necessary for celebration'. And so Skye experienced 'the innovation of a dry New Year'; dry, that is, in the sense that there was little to be had in the way of whisky. The weather was as damp as it was chill. And 'the prospect of a mild winter' was generally thought to be 'out of the question'.[1]

A fortnight into January it began to snow. Soon Skye's roads – many of them, in any case, little better than tracks – were blocked by drifts of 20 feet or more. 'For some days,' Gillies noted, 'the local merchants found it impossible to make contact with outlying townships and at times the shortage of rations gave concern.' A number of localities had to be supplied by sea. Livestock suffered extensively. Many sheep were smothered. And the local crofting population's stocks of winter fodder, seldom very plentiful to begin with, were alarmingly depleted.

Not since 1881, people said, had there been a blizzard of such severity. And the few elderly folk who could remember that earlier occasion – when the times had been so hard that a charitable collection, amounting to a famine relief fund, for the crofting areas had been launched in a number of southern cities – no doubt made the most of the temporarily heightened interest in their recollections of their youth. One such was Donald Finlayson from Gedintailor, a crofting township in the part of Skye that is known as Braes.

In 1945 Donald Finlayson was well into his eighties. Indeed he was to die that spring, just two or three weeks after the end of Hitler's war in Europe. And with Donald Finlayson's passing there was broken one

23

of Skye's few remaining links not just with 'the year of the great snow', as 1881 was remembered locally, but with the still more memorable sequence of events which had begun in Braes during the following winter and in which Donald Finlayson – 'a well known Land Leaguer', as he was described by Donald Gillies at the time of his death – had been an enthusiastic participant.

In 1881 the Braes people had initiated the rebellion which ensured the survival of crofting. In November that year, men like the young Donald Finlayson had marched from Braes to Portree, some eight or nine miles distant, to inform the local representative of their landlord, Lord Macdonald of Sleat, that no Braes crofter would pay the rent due on his holding until such time as there was restored to the Braes townships various traditional grazing rights of which the Macdonald estate management had previously deprived them. When the authorities eventually intervened in the ensuing dispute, on Lord Macdonald's side, men like Donald Finlayson – and women, too – fought savagely in the rain against the 50 or more policemen sent to arrest their leaders.

From that battle on the outskirts of Gedintailor in April 1882 a great deal followed. Soon crofters elsewhere in the Highlands and Islands were adopting the tactics pioneered in Braes. Rent strikes became endemic. Land was seized illegally. Troops were sent to Skye, Tiree and Lewis. But crofters, who had once seemed so helpless in the face of the landlords who had eradicated so many settlements during the mass-evictions or clearances which had been such a persistent feature of the nineteenth century in the north of Scotland, now refused to be cowed. Instead they organised themselves politically, joining the Highland Land League and pressing, by every available means, for the reforms which the Land League so vociferously demanded.

At last, in 1886, the British government, which had previously been slow to make concessions to crofting opinion, spectacularly gave way. That year there was passed a Crofters Act. And though the Highland Land League – in the manner of every group which has ever campaigned for radical change – inevitably wanted Parliament to do much more, it steadily became apparent that what had been done was remarkable enough. The freedom of landowners to do as they pleased with their own property had been drastically circumscribed. The prospects of crofters had been correspondingly enhanced.

No longer could a crofter be removed from his holding at his landlord's whim; no longer was he obliged to abandon all claim to his own investment in his croft on the day that, for one reason or another, he chose, or was forced, to relinquish it. Now a crofter had a legally guaranteed right to security of tenure for his own lifetime and, even more startlingly, an equal right to bequeath his tenancy to a family successor. Now croft rents were removed by law from the sphere of private contract and subjected to the judicial scrutiny of the tribunal which began life, in 1886, as the Crofters Commission and which, in 1912, was transformed into the Scottish Land Court. Now a landed proprietor was required to make appropriate financial recompense to an outgoing crofter in respect of any 'permanent improvements', including the provision of a house or any other building, which the crofter might have made to the holding he was leaving.

These changes, though fundamental, did not, of course, produce an overnight improvement in the position of each and every crofting family. More than 20 years after the passing of the Crofters Act, for example, the Poor Law Commission had occasion to compile a report on the Lewis parish of Barvas where, it was revealed, over 100 houses were 'glaringly and shockingly defective from a sanitary point of view'. In those gruesomely disadvantaged homes, commented the Commission, 'human beings, cattle and other livestock are all housed under the same roof without any effective partition wall; all enter by the same door as a rule and breathe the same air night and day, while the excretal matters of man and beast, and refuse and slops, are allowed to accumulate in the byre end of the house . . . . Drainage is almost entirely neglected about these houses and liquid sewage is permitted to find its way where it may.'[2]

But by 1909, when those words were written, such conditions were at least sufficiently exceptional as to be noteworthy. Prior to 1886 – when to improve a croft house was to lay oneself open to a steeply increased rent, maybe even to eviction if the estate manager, or factor, was looking for an above-average holding for a tenant to whom he owed a favour – homes of the type encountered by the Poor Law Commission in Barvas constituted the general rule throughout the crofting areas. Their gradual disappearance was one of the more readily observable indicators of the marked progress made as a result of the Crofters Act.

This improvement in housing standards, which was evident in most crofting districts well before the close of the nineteenth century, became still more rapid when the Board of Agriculture for Scotland was empowered, on its establishment in 1912, to make loans to crofters for housing purposes. A crofting tenant wishing to provide himself with a new home was now eligible for a Board advance of up to £150, repayable over 11 years. By 1939 almost £500,000 had been made available to crofters in this way. And almost 4,000 new croft homes had been constructed.

The necessary building work was mostly undertaken by crofters themselves. It was further assisted when the Department of Agriculture for Scotland, as the former Board was designated in the 1920s, agreed to establish a number of 'crofter stores' where materials such as cement and timber could be bought at competitive prices. It was little wonder, in such circumstances, that one visitor to the Hebrides in the 1930s should have commented that 'the arrangements made by the Department of Agriculture' with regard to housing were the subject of 'general praise among the people'.[3]

Equally deserving of approbation was the Highlands and Islands Medical Service Fund. Established by Herbert Asquith's Liberal administration in 1913, and in certain respects a regional precursor of the National Health Service which was to be introduced by Clement Attlee's Labour government in 1948, the Medical Service Fund ensured that no crofting family, however remote their home and however modest their financial resources, need be without access to a doctor at times of emergency. General practitioners were guaranteed a reasonable professional income and had their travelling expenses reimbursed. Hospitals were built. A district nursing service was founded. Although much remained to be done, even in the 1940s, there is no doubt that the Medical Service Fund, as one authoritative commentator noted, 'captured the imagination of all concerned' and made illness much less of a domestic catastrophe than it had previously been.[4]

The introduction of old age pensions in 1909 had similarly beneficial consequences. The cumulative effect of all these developments was such as to allow those Scottish Office civil servants responsible for Highlands and Islands administration to conclude that there had been, by the 1930s, 'a considerable evolution' in the overall condition of the north of Scotland. No longer were crofting communities characterised by the 'extreme poverty' which had been commonplace some 50 or 60 years before.[5]

Then the life of the typical crofter had been almost unimaginably precarious. He was constantly at risk, as was proved all too conclusively in 1881, from the hunger and want which necessarily followed the loss of his crops or the failure of his fishing. He might be compelled to remove himself, with no more than a few weeks' warning, from his home, his land, his township, his locality. Being invariably poor, usually illiterate, having no vote and being unable, very often, even to converse in the language of his country's rulers, he was singularly incapable – prior to his mobilisation by the Highland Land League – of forcing his plight on the attention of either the Government or society at large.

Now, within the lifetime of a man like Donald Finlayson, crofting prospects had become enormously less uncertain. The snowstorm which blocked Skye's roads in January 1945 might have caused considerable inconvenience. But it did not precipitate anything like the general food shortage experienced by islanders in 1881. And what was true of the results of bad weather was true, too, of the impact of other hazards. The threat of eviction had been removed entirely. Even sickness, old age or disability were no longer quite so frightening as they had been when the victims of misfortune were forced to depend completely on the charity of friends and relatives scarcely better placed financially than they were themselves.

While they would have readily acknowledged that some part of their progress was due to their inclusion in national programmes of reform, of which the provision of a measure of state-funded social welfare was one example and the introduction of universal education was another, crofters were naturally inclined, in any discussion as to why things had improved, to attribute rather more significance to the existence, since 1886, of a body of laws and a set of institutions which were designed specifically to safeguard crofting and which were without parallel elsewhere in Britain.

It was all very well for the Scottish Land Court, grumbled one Shetland landlord, to go about the Highlands and Islands 'enjoying themselves hugely' and 'being philanthropic at other people's expense'. But none of his crofting tenants would have shared that particular gentleman's disenchantment – however understandable it may have been from the landowning point of view – with the organisation which, more than any other, was symbolic of their newfound security. 'No institution is more popular in the Highlands than the Land Court,' remarked one sensitive observer of the 1930s crofting scene; 'and deservedly so for it has discharged its delicate duties with conspicuous success.'[6]

Croft rents, which had once verged on the extortionate, became increasingly nominal as the Land Court inexorably enforced the principle, enshrined in the Crofters Act of 1886, that the sum paid annually by a crofting tenant to his landlord should not incorporate any element in respect of such value as might have been added to the crofter' holding by the crofter himself. In practically every instance, so the Land Court ruled, the proprietors of Highlands and Islands estates had originally provided crofters with nothing more than scanty patches of largely virgin territory. The croft house was the work of the crofter. So, too, were any byres, steadings, sheds, fences and footpaths. No landlord, it was now decreed, was entitled to profit, by way of increased rent, from these, or any other, results of his crofting tenant's labours.

As croft rents fell, so the crofter's disposable income tended to increase; partly because a much smaller share of the average crofting family's budget was now being appropriated annually by the family's landlord; partly because higher money incomes were becoming a little easier to obtain.

Few crofters had ever been able to rely entirely on their land to provide them with a livelihood. Soils were too infertile, the climate too adverse and holdings far too small to make it possible for the typical crofting tenant – particularly if his holding was one of the diminutive plots which were, and remain, characteristic of crofting in places like Shetland, the Western Isles and much of the West Highland mainland – to aspire to be a full-time agriculturalist.

When conditions were at their most appalling, notably during the first half of the nineteenth century, sheer force of circumstance had compelled crofters to cultivate every scrap of available land. But when paid employment was more readily to be had, and when cheap, imported foodstuffs began to appear in even the remoter localities, as they did increasingly in the years around 1900, it no longer made sense to keep pushing the frontiers of cultivation still higher up the hillside, still further out on to the moor. The effort previously expended on endeavouring to get crops to grow on boggy, rocky ground could be more profitably invested in the business of earning the cash needed to purchase first flour, then baker's bread and groceries, then oilskin coats and rubber boots, dressed timber, manufactured furnishings and all the other goods which rising living standards, and consequently enhanced expectations, were making it important for the crofter to acquire.

As a money economy took hold, both families and townships became less self-sufficient, more fully integrated with the world outside. More trips were made to other places. More and more items were imported. Traditional crafts like tailoring and shoemaking began to decline. Homespun clothes went the way of thatched roofs. There was jam, sugar, tea, bacon, canned beef, scones and biscuits where once there had been only oatmeal, herrings, potatoes and milk. And though the new diet was not necessarily more nutritious than the old, there was little doubt as to which one was preferred.

Now the croft was less a source of supplies for the table, more a provider of hard cash. Livestock began to take the place of food crops like potatoes – with more sheep and more cattle being reared for the market. The money thus raised was supplemented by the crofter and his family taking jobs of every kind, jobs which, ideally, could be done while continuing to live on the croft, but which, in all too many instances, entailed prolonged absences from home.

Women from crofting townships worked as domestic servants in nearby shooting lodges and faraway city suburbs. Other women, especially those from the islands, went each summer and autumn to east coast fishing ports, from Wick and Peterhead to Lowestoft and Great Yarmouth, to gut, clean and cure the countless millions of herring which were afterwards shipped to Germany and Russia.

Men, too, found employment in the fishing industry; working as hired hands on other people's boats in deeper waters; taking their own share of the inshore catch. Also at sea, for much lengthier periods, were the growing numbers of crofters, and crofters' sons, who served in the merchant navy or voyaged to the Antarctic with the whaling fleet. The crofter, it seemed, would tackle anything, travel anywhere. He was harvest-hand, navvy, roadman, ghillie, stalker, ploughman, shepherd, military reservist, joiner, mason, motor mechanic, railway porter and postman. He was the self-employed weaver who made it possible for the tweed industry to flourish in the Western Isles. And he was the paid labourer who planted the first of the endless conifers with which the

Forestry Commission began, in the 1920s, to transform the appearance of so many Highland glens.

Many crofters still spoke Gaelic. But few now did not speak English also. And in their overall outlook, as well as in their experience of work and wages, the crofting population were no longer a people apart. The way they lived was not, of course, the way that their fellow citizens lived in Glasgow, Manchester or London, but the differences were clearly much less than they had been.

The merchants' vans which, by the 1930s, called daily or weekly in most crofting townships supplied their customers with products that were, for the most part, indistinguishable from those on sale in any industrial centre. What the van, or the local shop, could not supply was bought by crofting housewives from mail order catalogues – those produced by Welsh firms like J. D. Williams were among the favourites – of a kind that were increasingly familiar throughout the United Kingdom.

The crofter of the 1930s smoked the same cigarettes, wore the same dungarees, read the same newspaper – usually the *Daily Express* – as millions of people in the south. He listened, on his newly acquired wireless set, to the same BBC programmes. He went about his local business, very often, on a mass-produced bicycle of the type used also by innumerable city clerks and factory workers. And if he was doing a good deal better than the average, the crofter of the 1930s, just like the more successful working man elsewhere, might well aspire to the ownership of a motorbike – or even a second-hand car.

'We find the crofter, more particularly the younger crofter, utterly bored with the romantic conception of his life and what the city dweller thinks good for him,' it was shrewdly noted by the organisation which took charge of crofting administration in the 1950s. 'He has long since rejected the role of the noble son of nature who rejoices in homely fare and draws strength from stern privation. He now wants above all to be a citizen of this country as others are, not a curiosity or a "character" . . . but a citizen . . . He prizes his Gaelic culture, but not to the extent of being treated as a museum-piece on account of it; he will assuredly prize it still more highly when it is no longer the bed-fellow of poverty and underprivilege.'[7]

But not everyone wanted the Highlands and Islands, where the first scheduled air services began operating in the 1930s and where motor buses now plied between most outlying areas and the nearest town, to become less isolated, less cut off from wider social trends. Indeed the dominant theme of most of the many books written about the region during the two inter-war decades was that it remained largely unaffected by the many changes then occurring in the rest of the country.

The bulk of these publications were overtly escapist, dedicated to the notion that the north of Scotland, and especially the Hebrides, were wholly unlike other parts of Britain. To dip into a typical example of the genre is immediately to enter a make-believe landscape which over-laps only rarely with the world that was actually inhabited by crofters and their families.

The man who most energetically and enthusiastically worked this particular literary seam was Alasdair Alpin MacGregor whose very name was redolent of the Celtic twilight which permeated his numerous books. To MacGregor – who lived himself in the south of England – a crofting community, as he informed readers of *Summer Days Among the Western Isles*, published in 1929, was no ordinary spot. Rather it was a place of 'terrestrial silence and celestial peace' where bumble bees were to be observed 'hurrying one knows not whither', where hills and isles were invariably 'faery-haunted' and where night brought not so much darkness as 'the magic peacefulness of dreamland'.

Inevitably and deservedly, Alasdair Alpin MacGregor turns up, only thinly disguised as 'the well known topographer of the Hebrides', Hector Hamish MacKay, in Compton MacKenzie's comic masterpiece, *Whisky Galore*, a novel which dissects island life more delicately, and certainly more skilfully, than anything which MacGregor or his imita-tors ever wrote. But though it proved difficult, even for so talented an author as MacKenzie, to parody successfully MacGregor's uniquely awful style, the original of Hector Hamish MacKay – who is to be found in *Whisky Galore* standing 'entranced' on an island beach where he is watching 'the placid ocean break gently upon the sand to dabble it with kisses' – was sufficiently goaded by MacKenzie's mockery to attempt to turn the tables on his more renowned tormentor.

The ingenious islanders who, in *Whisky Galore*, neatly circumvent all officialdom's efforts to deny them access to the limitless supplies of whisky in the holds of a wrecked ship sent providentially to relieve a wartime thirst of the kind that prevented Skyemen from doing proper justice to the start of 1945, become, in Alasdair Alpin MacGregor's ostensibly factual publication, *The Western Isles*, which appeared in 1949, a drunken, brawling, violent crew who appear to set aside their 'indolent and dilatory' habits only in order to indulge their alleged fondness for the wilder variants of sexual immorality.

The astringent MacGregor of 1949, however, was arguably as much at odds with reality as the sickly-sweet MacGregor of 20 years before. The prosaic truth, perhaps, was that crofters, in general, merited neither idealisation nor vilification; just as it was the case that the places where crofters lived could produce both days of sheer beauty and days when it was, quite plainly, hellish to have to be out and about. The naturalist Seton Gordon – whose contribution to the Hebridean literary industry probably excelled even MacGregor's but who was much less romantically inclined – experienced one of these stormy days in the Uists in the 1920s:

> Along the coast, a mountainous sea is running. On the western horizon enormous waves are breaking upon the rocky isle of Haskeir and their drifted spray rises to a height of fully a hundred feet. Around the . . . crofting village of Houghary there is much activity. During the summer, bad weather has been almost continuous and now it is November and the oats, rye and barley are still unsecured . . . . Already the crofters have lost a portion of their crop, for a fierce gale lifted the sheaves and blew some of them into the Atlantic. And so the men and women of the village are now busy carrying large stones into the fields to hold down their sodden harvest. With great labour the scattered sheaves are collected and are tied together with ropes weighted with stones. But some of the crop has been blown so far that it lies in pools of water, or upon the road, and it is impossible to tell who is the owner of these stray sheaves.[8]

Despite such hardships, of course, the crofting life was not without its compensations. And crofters, had they been in the habit of analysing their own feelings in such terms, which most of them were not, might well have produced judgments on crofting that were not too far removed from that actually delivered by Raymond O'Malley, one of the first of what was many years later to be a growing band of southerners attracted to the Highlands by the prospect of taking over a croft.

There was a particular 'satisfaction' to be had from crofting, concluded O'Malley, who moved to a holding at Achbeg, near Stromeferry,

in 1943. There was, he elaborated, 'the variety of work, the co-operation with natural forces, the struggle with the weather, the sense of
difficulty overcome, the intimate acquaintance with wild creatures and
with natural beauty, the mutual help of neighbours . . . the feeling of
community.'[9] But crofting, as O'Malley soon discovered, and as any
practising crofter could have told him at the outset, had its drawbacks
as well as its pleasures.

> There was the aching vastness of weedy potato drills, or of spread hay with
> a shower racing in from Applecross; the worry of March and April as the
> stacks dwindled and still the grass showed no sprout of new green; the sordid
> Sunday cackle of the hens; the trickle down the spine and squelch of feet
> in leaky boots; the endless drizzle as the grass went to stem and seed; there
> was maggot (in the sheep), that living death; the weeks of painful waiting
> for the dentist; wartime whisky; and the midges, the petty, crawling,
> stabbing, insignificant, insufferable multitudes of midges.

But the relationship between a crofter and his croft was no simple
thing, no straightforward weighing of costs against benefits. Here is one
Lewis crofter, Robert MacLeod from Carloway, on that relationship's
manifold complexities.

'In the family in which I grew up,' MacLeod wrote in the early 1950s,
recalling a period nearly half a century before, 'as each one came to the
stage where criticism is voiced as well as complaint . . . we asked our
parents to get rid of the croft, house and all, and clear out. There was
no special preference where to, except that it would undoubtedly have
to be a place where neither children nor their mother carried creels of
peat and pails of water; where sooty rain never dripped through the roof
and where people had never heard of a manure heap.'[10]

But the repugnance with which crofting was often regarded by
crofters themselves did not inevitably result in holdings actually being
abandoned. Running counter to those negative feelings and impulses
which he had already described, Robert MacLeod explained, was 'a
strong bond of affection' for the croft – a bond, as the Carloway crofter
put it, which owed more to 'instinct' than to 'rational thought'. And
that sense of attachment to the land, MacLeod stressed, completely
transcended all the many legal formulations concerning the distinction
between tenancy and ownership.

'As a crofter,' Robert MacLeod observed, 'I have great difficulty in conceding to the proprietor any greater right to the croft than I have myself. It is the four small acres my family broke in from the wild morass and rock, drained, manured and fed, over the years, with no help from the proprietor . . . . I hold this individual piece of land to be mine morally and to be disposed of as I wish.'

This observation, as was concluded by the enquirers at whose request MacLeod set out his views, captured the common crofting attitude: 'Above all they have the feeling that the croft, its land and its house, are their own. They have gathered its stones and reared its buildings and occupied it as their own all their days. They have received it from their ancestors who won it from the wilderness and they cherish the hope that they will transmit it to the generations to come. Whatever be the legal theory, they feel it to be theirs – and, in this respect, the provisions of the Crofters Acts do no more than set the seal of Parliamentary approval on their deepest convictions.'[11]

Such sentiments, perhaps, were more self-evidently admirable to outside observers than they were to crofters themselves. To many a person brought up on a north of Scotland smallholding, the pull exerted by the croft seemed as much of a curse as a blessing. 'If I were not born there and the very dust of the place dear to me,' said one crofter in a moment of exasperation with his fate, 'I would quit tomorrow.'[12]

As it was, that man stayed home to make the best of things. And what could be made of one's life on a croft was often quite extraordinary – as is demonstrated by this summary, made in 1952, of the accomplishments of Archibald MacMillan, who had taken over a six-acre holding in Ardnamurchan some 20 years earlier.

> The land was then in good order but the buildings were practically useless. He built the dwelling house and steading . . . . He brought up a family of four boys and two girls. His eldest son was finishing his veterinary course in Glasgow. His eldest daughter was a nurse and his second daughter was teaching in his own area. He found that his croft would sustain himself and his family for about five months of the year. For the rest of the year he took all the work he could get in the district – manual work and haulage work with a horse and cart. If he had not had that work, he could not have carried on.[13]

Archibald MacMillan's reliance on money made from activities other than those which had to do with cultivating his land was, as already mentioned, wholly characteristic of crofting. No matter how intensively a crofter managed his holding, the financial return on his crofting labour was almost always inadequate to meet his needs. And that return was, if anything, inclined to fall, and sometimes to fall steeply, throughout the 1920s and 1930s – years when agriculture, in the United Kingdom as a whole, was doing far from well.

Even in England's most fertile farming areas, in the course of those bleak and hopeless decades, 'millions of acres' were said to have 'passed out of cultivation'. Farm buildings, it was reported, were in a 'deplorable' condition. And farmers were as poorly off as they had ever been. 'Field after field may be seen with hideous crops of weeds, reeds, thistles, nettles and brambles. The stock is very scanty and the gates and fences . . . are in disrepair.'[14]

Agricultural prices, which had benefited from the protected market established as an emergency measure during the Great War, had plummeted in the 1920s when free trade in food products was restored. Although hill sheep producers, crofters among them, had profited initially from the demand for breeding animals on the part of those south country farmers who were taking their land out of cereals in order to turn it over to livestock, sheep prices, too, eventually collapsed. Between 1930 and 1931 the prices fetched in Scottish markets fell by some 40 per cent, with an even sharper decline in 1932. Blackface lambs, the major crofting product, were selling for no more than a few shillings each. At one auction ring in Sutherland, during that 'period of almost unbroken depression', as the thirties were dubbed in a subsequent farming survey, the only offer made for one batch of small lambs amounted to sixpence a head.[15]

Hardest hit by the agricultural slump, paradoxically, were those crofters who had been the principal beneficiaries of previous land reforms. In the years around 1920, in response to the crofting community's persistent demands for the restoration of lands lost in the course of the clearances of 50 or 100 years before, the Government had embarked on an ambitious programme of land settlement. Throughout the Highlands and Islands, but most particularly in the Hebrides, the farms from which a previous generation of crofters had been forcibly removed were bought by the state and used to create many hundreds of new crofts. But to be allocated one of those publicly provided holdings was not, as things turned out, to be so favoured as first it must have seemed.

The farms purchased by the Government, in the guise of the Board – or Department – of Agriculture, were substantial holdings with large sheep flocks. The Department was required to buy those flocks from the outgoing farmers at a time when, in the immediate aftermath of the wartime farming boom, prices were still extremely high. And in order to recoup its own initial outlays, the Department was obliged, in turn, to insist that newly settled crofters, whose landlord the Department now became, took the former farm sheep stocks off its hands. Loans were made available to the Department's crofting tenants for this purpose. But by the early 1930s, when virtually every sheep production business in the country was failing to break even, let alone turn in a profit, crofters were finding it virtually impossible to meet the Department's repayment demands.

Eventually, in the face of protest meetings and petitions, the Department of Agriculture temporarily suspended the interest charges due on its tenantry's outstanding sheep stock debts. That brought some small relief. But no interest charge rebatement, in itself, could be sufficient to make a worthwhile impact on a crofting crisis which extended far beyond the Department's land settlement estates.

Crofting families of the 1920s and 1930s, as was emphasised in the first part of this chapter, were much better placed, in practically every way, than were their nineteenth-century predecessors. But that very state of affairs, as was also stressed, naturally helped to make crofters much less inward-looking, more inclined to make comparisons between their own position and that of people living in other parts of Britain. Relative to much of the rest of the country, it was clear to anyone who read a newspaper, the Highlands and Islands remained very far from prosperous.

This was a situation neatly calculated, as was observed by one social scientist who made a close study of the crofting areas in the 1930s, to produce strongly contending emotions in the crofter. 'He is urged in opposite directions by his instinctive desire to live on in the place where his race and family were cradled and by his desire to share in the materially richer life he sees in other parts of the country.'[16]

Crofters like Robert MacLeod, whose views have already been quoted, resolved this dilemma by electing to remain on the holdings

they inherited. But growing numbers, from among the younger generation in particular, made an opposite decision. It was wrong to blame
them for so doing, remarked Colin Macdonald, who had himself grown
up on an Easter Ross smallholding and who, initially as a civil servant
in the Department of Agriculture and latterly as a member of the Scottish Land Court, developed an unrivalled knowledge of the crofting
scene. 'It is unreasonable,' Macdonald commented, 'to expect educated
young people to stay contented and happy in an unremunerative calling
while bigger prizes are offered elsewhere.'[17]

Thousands left Lewis alone in the 1920s. If the exodus from other
crofting localities was less spectacular, that was mainly because their
people had begun to go away much earlier. Remoter islands, of which
St Kilda was, of course, the most heavily publicised example, were
abandoned altogether. Soon there was talk of the depopulation process,
in the more isolated parts of the mainland also, having reached the
point where it was 'endangering normal social life'.[18]

That this was no exaggeration was given dramatic confirmation on the
outbreak of war in September 1939. In the crofting parish of Gairloch,
it was afterwards reported, there were then only 30 men of military age
– less than a quarter of the number who had volunteered for service with
the Gordon Highlanders on a single day some 25 years before.[19]

There was little dispute as to the causes of the outflow. 'Education,
higher standards of living and of health, contact with industrial activity
and with town life, have brought with them the natural desire of the
rising generation to share in the general social advance.' So concluded
one more of the many official committees formed to look into Highland
problems.

In order to 'share in the general social advance', it was widely agreed,
a younger person from a crofting background was virtually compelled
to leave home; not just because the typical crofting district was notori-
ously deficient in every sort of social amenity from cinemas to pubs and
dance halls; not just because it could be anything but pleasant to grow
up in communities which were becoming increasingly aged in character
and increasingly repressive in outlook; but because, above all, there was
not sufficient well-paid work on offer in the Highlands and Islands to
permit the region's younger people to secure the sort of living standards
to which more and more of them thought it perfectly legitimate to
aspire.[20]

37

One group of Scottish trades unionists observed:

The days of a crofter living off his land are past, and not all the romantics in the Lowlands will make them return. It is from this understanding that the problem of depopulation can best be approached. People leave the High-lands because a reasonable living cannot be made in them. This may appear rather an obvious truth, but it cannot be overemphasised. It is too often said that the reasons for depopulation are the lure of the city lights, the poor housing, the lack of electric light or the poor transport. It is true that the lack of these things may drive people to the towns. But the lack itself is only a symptom of the main cause – poverty.[21]

Poverty was becoming steadily more apparent in the north of Scotland in the 1930s. The decade following the Wall Street crash of 1929 was one of worldwide depression. And the Highlands and Islands were by no means immune from the debilitating consequences of economic dislocation and collapse.

Men and women who had left their native townships, as they thought for ever, now returned. Even a Hebridean croft, it seemed, was a better place to be in these grim years than was the line outside a Detroit soup kitchen. But the difference was one of degree only. To have no resources except those provided by the family croft was not to be at all well placed. More and more folk, however, found themselves in exactly that unenviable position.

Ship after ship was tied up on the Clyde and on the Mersey. Seaman after seaman was sent home to Lewis, Uist, Skye or Shetland where each such refugee found himself in the same inescapable predicament as the equally numerous victims of the parallel contraction of the fishing industry. By 1938 there were only one third as many men employed in fishing off the north of Scotland as there had been in 1913. Practically every other trade of any consequence was on a similar downward slide.

In 1936 the unemployment rate in the Highlands and Islands was officially stated to be just under 35 per cent. The number out of work had more than doubled since 1927. Money was in such short supply that some crofting communities showed signs of reverting to the cashless conditions of the distant past.

The greatest asset of the unscrupulous local dealer was now his crofting customer's indebtedness. The crofter who came to plead with a Stornoway or Lerwick merchant for his household provisions would be offered the goods he needed in return for payment in kind. But what was supplied to the crofter might well go down on the dealer's ledger at the highest retail price. And what was accepted from the crofter – some eggs, perhaps, a hen or two, a hundredweight of potatoes, even a calf or half a dozen lambs – was often entered at the lowest possible wholesale figure. The bad old days of truck and barter had returned.

But Highlands and Islands problems were not sufficiently spectacular to have secured worthwhile Parliamentary attention, wrote David Keir, a *News Chronicle* correspondent who conducted his own fact-finding mission in the north of Scotland in 1938. 'Jarrow, Ebbw Vale, the Rhonnda forced themselves on the public conscience,' Keir continued. Their 'derelict works', photographed repeatedly 'as a background for inquiring politicians', had 'hit the front page so often and aroused such public outcry' that the Government had been 'forced' to introduce its so-called Special Areas legislation in an attempt to assist their regeneration. 'But the laying up of fishing boats, the spread of bracken and the gradual departure of the Highland population,' Keir remarked, 'is less dramatic, less noticeable.' And so, he concluded, national politicians would go on ignoring what was happening in the crofting areas.[22] But there David Keir was wrong.

The House of Lords has had a longstanding interest in Highland issues. Much of the north of Scotland, after all, has traditionally belonged to its members. As more than one speaker indicated, when the Upper House once again turned its collective attention to Highland affairs on 13 July 1939, even those peers who were unlucky enough not to have a grouse moor or deer forest of their own were generally in the habit of spending their summers on the Scottish estates of other noblemen more fortunately situated.

So it was not really surprising, even at a time when Nazi Germany's aggressive intentions on Poland were becoming daily more apparent, that the House of Lords should have been discussing the Highlands and Islands. What was rather more out of character – given the aristocracy's habit, dating back to the debates surrounding the passage of the 1886 Crofters Act, of doing its level best to obstruct legislative initiatives intended to promote crofting betterment – was the passionately concerned tone of the peerage's comments on the many difficulties then confronting crofters.

'It is no small problem that we are discussing,' declared the Earl of Elgin in his opening speech, 'but a problem affecting almost exactly half the area of Scotland.' In that wide swathe of territory, the earl continued, the total population had 'diminished by one third' in the course of just over half a century. 'The schools are emptying, the young men and young women are leaving, the crofts are being deserted and neglected, the produce of the land is decreasing and occupation on the sea, and livelihood from it, is collapsing.'[23]

One 'typical example' of a 'formerly thriving' community, Elgin thought, would bring the scale of the problem 'forcibly before' his listeners. 'In a small village on the west coast of Skye, there were, 50 years ago, 100 persons.' Now that settlement's entire population numbered just 19 – and five of those were old age pensioners. 'For four years,' Elgin commented, 'there has not been a child in the school.'

This, he continued, constituted 'a picture of sadness and decay'. Elgin, for one, was in no doubt as to what had to be done. 'We cannot afford to let this heritage of ours in the Highlands and Islands dwindle away before our eyes and I urge the Government to open their eyes as well as their hearts and overcome evil with good.'

That the Government ought to be dealing more constructively with crofting problems was a view held also by John Colville, the west of Scotland industrialist who then held the post of Scottish Secretary in Neville Chamberlain's cabinet. It was 'essential', Colville had informed a somewhat sceptical Treasury Minister a month or so earlier, that substantial funds should promptly be found for the Highlands and Islands. Public opinion demanded as much, the Scottish Secretary stated flatly. And the House of Lords debate, he added in the course of a later exchange of correspondence, had provided an 'indication of its force'.[24]

Such a strongly interventionist stance seems strangely at odds with the attitudes of many more recent Conservative politicians. But Colville – who was, incidentally, in the habit of holidaying in the Hebrides and who consequently knew the crofting areas fairly well – was no maverick in this respect. Only if the Scottish Office was prepared to get involved directly in the promotion of economic expansion, it was believed, not just by Colville but also by his immediate predecessors, Sir Godfrey Collins and Walter Elliot, could the Tory-dominated national government, which had held office since 1931, maintain its grip on Scotland in the face of the electoral threats posed by both Labour and the then newly formed Scottish National Party.

Nor were such threats confined to the Central Belt. At the general election of 1935, the national government had lost the Western Isles to Labour. Not only had this predominantly crofting constituency been taken from the ruling administration by a 22-year-old student, Malcolm MacMillan, a man of markedly left-wing convictions; the SNP share of the vote, at just under 30 per cent, had also been far higher than anyone had anticipated.

The 'highly probable' result of this particular debacle, observed the Government's defeated candidate, T. B. Wilson Ramsay, in a letter to a former colleague, was that the 'whole Highlands' would eventually fall to Labour and the Nationalists. Only if the Scottish Office were to take positive measures to combat the economic crisis affecting all of northern Scotland, Wilson Ramsay continued, could such an outcome be forestalled. What was needed, in his opinion, was nothing less than the immediate establishment of a 'Highlands and Islands Development Fund' to which the Government ought to be prepared to make 'a contribution of, say, one million pounds a year'.[25]

At its annual conference in Dundee in the November following the 1935 election, the Scottish Conservative and Unionist Association – in a highly unusual display of independence – carried a similar proposal in the face of more cautious noises from what *The Scotsman* called 'the official side'.[26] From people who were outside the ranks of the Tory Party, needless to say, there were much more far-reaching demands.

In a pamphlet published in the autumn of 1935 and entitled *A New Deal for the Highlands*, a Ballachulish doctor, Lachlan Grant, pointed to the Roosevelt administration's massive developmental efforts in the more rural parts of the United States and sharply contrasted this American activity with the British Government's comparative neglect of the northern half of Scotland.

Soon Grant was chairman of the Highland Development League which was launched in Glasgow in February 1936 and which, it was hoped by its founders – prominent among whom was the Gaelic-speaking Liberal activist John M. Bannerman – would quickly put the

41

Government under pressure of the type generated so effectively by the Highland Land League in the course of earlier reform battles. But such pressure proved to be unnecessary. The Development League, it quickly became apparent, was charging at a door which the Secretary of State for Scotland, Sir Godfrey Collins, was already holding open.

Not the least commercially successful representative of the Glasgow-based publishing empire of William Collins and Sons, Sir Godfrey was determined to reinvigorate the recession-scarred Scottish economy by means of agencies such as the Scottish National Development Council which was headed by another Glasgow businessman, Sir James Lithgow, and which also had the support of organised labour in the person of William Elger, general secretary of the Scottish Trades Union Congress.

At the Scottish Secretary's instigation, the National Development Council had established, in March 1936, a Scottish Economic Committee. This committee, Collins announced in April, would 'examine the possibility of improving economic conditions in Scotland'. In the course of the following summer, again with the Secretary of State's encouragement, the Economic Committee set up a further specialist grouping with the remit of enquiring into 'the economic conditions of the Highlands and Islands'.[27]

The Highlands Sub-Committee of the Economic Committee of the Scottish National Development Council – to give the new body its full, if more than somewhat clumsy, title – was chaired by a Skye landowner, Major Edward Hilleary, a former finance convener of Inverness-shire County Council. Among its members were Sir John Sutherland of the Forestry Commission, Murray Morrison of the British Aluminium Company, which had established smelting operations at Kinlochleven and Fort William, Flora MacLeod of MacLeod, like Hilleary a crofting landlord, and Joseph Duncan of the Scottish Farm Servants Union.

Although the Hilleary Committee was, strictly speaking, an entirely unofficial organisation, its relations with the Government were both close and cordial. The committee, Ministers agreed, should have its expenses met from public funds. And by way of further encouragement to treat their job with all due seriousness, Hilleary and his colleagues were informed by John Colville, then Parliamentary Under Secretary of State at the Scottish Office, that the Government was 'confident that the committee could produce a report that would give both inspiration and guidance in the solution of a great human and economic problem vital to Scotland's welfare'.[28]

In December 1936 a similar message was conveyed to MPs by Walter Elliot who had become Scottish Secretary in succession to Collins and who was, if anything, even more committed to the notion that the state should have a key role in northern Scotland's economic and social regeneration.

The occasion was a debate at the end of which the House of Commons adopted a motion that, 'in view of the widespread and con-' tinuing distress in the Highlands and Islands of Scotland now involving grave peril of a complete breakdown in the social economy of these areas', the Government should 'formulate without delay proposals designed to arrest depopulation and poverty among the remaining inhabitants and to provide facilities whereby they may earn a decent livelihood'. And this, it seemed, was a task which Walter Elliot was perfectly happy to take in hand.[29]

The son of a Lanarkshire cattle auctioneer and the grandson of a Highland sheep farmer, Elliot was one of those Conservatives who was anxious to avoid 'the polarisation of our people', as he put it, 'into the haves and the have-nots'. As Minister of Agriculture prior to his move to the Scottish Office, Elliot had responded to the countrywide slump in farm prices by making available to the agricultural industry the first of the many financial supports and subsidies on which farmers were to become increasingly dependent in the years ahead. And now, as Secretary of State for Scotland, Walter Elliot clearly saw no reason to refrain from similar state intervention in the north of Scotland.

He very much hoped that the National Development Council's Highlands Sub-Committee would 'bring forward far-reaching proposals', the Scottish Secretary told MPs. And if such proposals were indeed to be forthcoming, they would certainly receive his 'careful and sympathetic' attention. 'The Government which has taken such vigorous and drastic action to maintain the great industry of agriculture,' Elliot continued, 'will not shrink from equally drastic action in extending and developing that and other industries if it is so recommended in a well thought out plan by the Highlands Sub-Committee.'

Within a day or two of the Secretary of State's House of Commons speech the Scottish press was carrying informed speculation to the effect that it was 'not improbable' that a 'special development board' would be 'established with wide powers to co-ordinate and give effect to schemes for the betterment of agriculture, fishing, afforestation, tourist traffic, transport and other industries in the Highlands and Islands'.[30]

Major Hilleary and his colleagues, meanwhile, were busy in the north where, one newspaper reported, they had been very much struck 'by the desolate state of affairs'. He had been utterly appalled by the conditions he had observed while travelling through the crofting areas, one member of the enquiry team publicly confirmed. He was accordingly determined that matters should be put to right. 'What we need nowadays is a new approach to the Highlands. We must clear our minds of all sentiment and cant. They should be regarded as a part of the country which has the same right to develop as other parts.'[31] When its report eventually appeared, in November 1938, it was evident that these were also the views of the Hilleary Committee as a whole.

On crofting, in particular, the committee's findings were categorical: 'If crofting is to take its place as a definite part of the general scheme of national life, it must be raised from being merely an existence to an occupation offering reasonable prospects of affording a living.'[32] To that end the Hilleary Report made a number of specific recommendations.

Agricultural training centres and model crofts, to be used for demonstration purposes, should be established throughout the crofting areas. Government subsidies should be given to shipping operators in order to reduce the freight rates which tended to render crofting products uncompetitive. And, most important, a Central Marketing Agency should be set up to provide 'facilities for transport of stock to market' as well as to make arrangements 'for the supply of farming requisites at co-operative prices'.

But these measures by themselves, Hilleary and his colleagues were clear, would not solve crofting problems. 'Two prominent facts have to be faced,' ran one of their report's more central passages, 'the first of which is that crofters . . . form the bulk of the Highland population and, second, that the majority of these men must find employment additional to their ordinary occupation in order to make a living. These facts lead to the decisive conclusion that a satisfactory solution to the difficulties through which these men are passing will not be found without the provision of work.'

To make available the jobs which were urgently required, observed the Hilleary Committee, would entail Government action on an unprecedented scale. 'The establishment of industries of a suitable character . . . becomes a necessity. The attempt must be made to expand those which now exist . . . and an organised plan for the introduction of others must be framed.'

The areas of economic activity which the Hilleary Report reckoned to have the greatest future potential were forestry, tourism and hydro-electric power. The institutional mechanism which the report suggested should be put in place to assist with their development was drawn from the National Government's Special Areas Act of 1934.

This was legislation which had been passed in an attempt to cope with the mass unemployment then blighting the urban centres most conspicuously affected by recession. In order to implement the 1934 Act's provisions in Scotland the Government had appointed a so-called Special Commissioner whose task it was to sponsor new industrial projects in places like West Lothian, Fife and Lanarkshire. Now, so Major Hilleary and his colleagues maintained at any rate, a second such Commissioner should be given responsibility for the economic well-being of the crofting counties.

Although there was strong backing in the north of Scotland for the suggestion that the Government should invest directly in the Highlands and Islands economy, Hilleary's recommendation that such investment should be managed solely by a Special Commissioner was less well received. Within a couple of months of the publication of his committee's report, the notion of a Highlands and Islands Development Board, of the kind which had been mentioned two or three years earlier, was back on the political agenda following the intervention of Sir Alexander MacEwen, an influential figure who, as Provost of Inverness in the early 1930s, had had a good deal to do with the formation of the Scottish National Development Council.

Originally Liberal in his politics, MacEwen had helped launch the SNP and had been that party's parliamentary candidate in the Western Isles in 1935. Now, in a series of speeches and publications, he dismissed the Special Commissioner concept as likely to result in 'virtual dictatorship' and pressed instead for the establishment of a four-man Development Board which, MacEwen stressed, ought ideally to be chaired by a Gaelic-speaking Highlander 'who would command general confidence'.[33]

At a private meeting in the Scottish Office in February 1939, John Colville, who had become Secretary of State some months before, confessed that 'his mind was still open' as between the 'two alternatives' of MacEwen's Development Board and Hilleary's Special Commissioner. But the obstacles in his way, Colville commented, were more financial than institutional. He was 'anxious that something should be done for the Highlands', the Scottish Secretary stressed. But the international situation was becoming steadily more threatening. It was consequently 'a difficult time at which to ask for more money from the Treasury in view of the nation's expenditure on its defence needs'.[34]

Within weeks, unfortunately, Hitler's march into Prague had made the times more difficult still. Such were the public expenditure implications of the massive rearmament programme now set in train that Colville was obliged to abandon all hope of either a Special Commissioner or a Highlands and Islands Development Board.

But so great was the interest aroused by the Hilleary Report, 'not only in the Highlands but in Scotland generally,' the Scottish Office informed the Treasury in June 1939, that the Highland issue could not simply be shelved. The Secretary of State, Treasury officials were told, 'recognises, of course, that what might be done in happier times is not practicable today'. But it was nevertheless 'essential' that 'some steps should be taken' to implement the Hilleary Committee's more important proposals. Consent for additional Scottish Office spending of £114,000 was accordingly sought. This would not go far towards meeting the wider developmental requirements of the north of Scotland. But it would, at least, enable some new roads to be built and some new fishing boats to be provided.[35]

There followed a protracted political wrangle. On the Treasury's offering considerably less than half the cash he had requested, an angry Colville responded with the comment that he would 'find it difficult to face the House of Commons' if his Highland policy was to be so circumscribed. But in the end, such was the Treasury's determination and so insistent were the demands of the defence departments, that the Scottish Secretary was obliged to announce, at the beginning of August, that his new spending in the Highlands and Islands would add up to no more than £65,000 in the year ahead.[36]

There was immediate uproar from MPs, local authorities and others. But what its eventual outcome would have been is, sadly, a matter for abstract speculation. Towards the end of the hefty clutch of Scottish Office files generated by the Hilleary Report and its aftermath is a letter written by one of Colville's Cabinet colleagues. It is dated 2 September 1939, the day following Adolf Hitler's invasion of Poland. And it simply comments, with regard to the Scottish Office's plans for the Highlands and Islands, 'that this question has now been swamped by more urgent matters'.[37]

On 18 September, a fortnight after Britain had gone to war with Nazi Germany, the bulk of such development spending as had already been sanctioned was formally cancelled. An early experiment in interventionist economics had come to a premature and inglorious conclusion.

But by senior politicians like Walter Elliot and John Colville the particular claims of the crofting areas on government had, at least, been recognised. And that recognition, even in time of war, was not again to be withdrawn entirely. There was no question now of government not taking an active interest in crofting issues. The only question to be settled was the form that such interest ought to take.

Chapter Two

NO ANSWER TO THE CROFTING PROBLEM

The outbreak of hostilities with Germany in September 1939 inevitably put a stop to talk of government action to expand the Highlands and Islands economy. But the massive military effort then embarked upon, ironically enough, was to produce just such expansion.

To one of the men who had been involved in the compilation of the Hilleary Report, it seemed by 1945 that the north of Scotland had been transformed even more than the rest of Britain by the impact of the war. 'The most obvious change was the absence of a very high proportion of the young men of fighting age in the services . . . and of the young women in the services and in the munitions industries. The second main difference was that there was more money in the Highlands than at any time in history.'[1]

A Government which, in the summer of 1939, had been unable to find more than an additional £65,000 annually for all of northern Scotland was soon afterwards spending much larger sums in single islands in a matter of months or even weeks. To counter Germany's highly successful U-boat campaign in the Atlantic, and to meet the strategic threat posed by the Nazi conquest of Norway, the Royal Air Force established major bases at several points in both the Hebrides and Northern Isles. The Royal Navy, meanwhile, was taking charge of convoy assembly points in West Highland sea lochs, and the army was stationing its troops throughout the region. The resulting demand for civilian labour was practically insatiable.

In Shetland there were times when mililtary personnel outnumbered the locally resident population. Both there and in Lewis, where men had to be transported daily by bus and truck from as far away as Uig and South Lochs to work on the construction of the Coastal Command aerodrome outside Stornoway, the level of manual earnings reached unprecedented heights. The Western Isles, it was commonly remarked, had never been so prosperous.

Much was done in these years which had previously been said to be impossible. Deer forests, against which successive generations of Highland land reformers had campaigned – for the most part unavailingly – on the basis that the millions of acres devoted to sport should be restored to agricultural use, were now compulsorily requisitioned and turned over to the rearing of livestock. Producers of hill sheep and hill cattle – people who had been told throughout the 1930s that they simply had to put up with low prices – suddenly found themselves the recipients of an entirely novel form of income which, it was hoped, would persuade them to turn out more of the lambs and calves which the country so badly needed.

On 12 December 1940 the Government announced that each hill ewe in Scotland would qualify, during 1941, for a support payment of two shillings and sixpence. That figure was to rise to eight shillings annually in 1943, by which time there was also a hill cattle subsidy of some two pounds a head.

Presiding over all such measures was a man who, it later came to be considered widely, was Scotland's most successful, and certainly most powerful, Secretary of State. Tom Johnston's political career had begun in Glasgow some 30 or 40 years earlier when, as an aggressively socialist journalist and pamphleteer, he had attracted a good deal of notoriety on account of the sheer ferocity of his attacks on northern Scotland's landlords – whom he held wholly responsible for all the many ills endured by crofters since the time of the clearances. Following his subsequent election as a Labour MP, Johnston, like many others both before and since, had mellowed a good deal. But his concern for the welfare of the crofting areas had remained as strong as ever. When Winston Churchill asked this prominent Labour politician to join his Coalition Government as Scottish Secretary, it was, at least partially, with an eye to clearing the way for new initiatives in the Highlands and Islands that Johnston accepted only on condition that he would be permitted to conduct policy in Scotland very much as he personally pleased.

Soon Johnston had constituted in Edinburgh what amounted to a governing council consisting of himself and five previous Secretaries of

State. Among the latter were men like Walter Elliot and John Colville who, as already mentioned, had been far from unwilling, even in the much less propitious circumstances of the 1930s, to get seriously to grips with the problems of the north of Scotland. Now they gave their unstinting backing to Tom Johnston's attempts to realise his lifelong ambition of reversing the depopulation of the Highlands and Islands. At the initial meetings of his council, therefore, the new Secretary of State had no difficulty in securing general agreement to the formation of committees of enquiry into several matters of considerable concern to northern Scotland.

One committee was to look into the state of the herring fishing industry. Another was to examine the possibility of a renewed programme of land settlement. There was to be an assessment of hill farming prospects. And from an especially impressive enquiry team headed by Lord Cooper, the Lord Justice Clerk, there was to be a report on 'the practicability and desirability of developments in the use of water power in Scotland for the generation of electricity'.

The Hilleary Committee had already urged that steps be taken to exploit the hydro-power potential of the Highlands. Now the Cooper Committee delivered the same verdict in a well-researched report that was quite scathing in its dismissal of those sporting and other interests which maintained that dams and power stations should not be permitted to impinge upon the north of Scotland landscape.

> If it is desired to preserve the natural features of the Highlands unchanged in all times coming for the benefit of those holidaymakers who wish to contemplate them in their natural state during the comparatively brief season imposed by climatic conditions, then the logical outcome of such an aesthetic policy would be to convert the greater part of the area into a national park and to sterilise it in perpetuity providing a few 'reservations' in which the dwindling remnants of the native population could for a time continue to reside until they eventually became extinct. But if, as we hope and believe, the policy, to which this report is a small contribution, is to give the Highlands and the Highlanders a future as well as a past and to provide opportunity in the Highlands for initiative, independence and industry, then we consider a few localised interferences with natural beauties would be an insignificant price to pay for the solid benefits which would be realised.[2]

These conclusions were very much the Scottish Secretary's own sentiments, and he moved at speed to implement Lord Cooper's findings. The committee's report appeared in November 1942. The Hydro-Electric Development (Scotland) Bill received its second reading in the House of Commons in February 1943, and the North of Scotland Hydro-Electric Board was an accomplished fact by the following summer.

The new organisation, MPs were told by Malcolm MacMillan – who had so startlingly taken the Western Isles for Labour in 1935 and who, in the course of his long parliamentary career was to be one of the most staunchly committed political advocates crofters have ever had – would help bring 'full employment, a sure livelihood and a decent standard of living' to the Highlands and Islands. 'It will be a blessing when we can supply the domestic user with electricity,' MacMillan continued, 'so that I shall see, before I leave this House, the time when it is not necessary for women in the Western Isles to go out cutting peat and carrying it home . . . . I hope to see the day when the people of the Western Isles will no longer have to use paraffin lamps, with all their dirtiness and expense and danger.'[3]

The cutting of peats was not, in the event, to be abandoned altogether. In other respects, however, MacMillan was to get his wish. One of the present author's earliest recollections of growing up in a small community in North Argyll in the 1950s is of the day that the 'electric lights' could at last be switched on and the oil-fuelled Tilley lamps put away for ever. While it was certainly the case that by no means all the more extravagant hopes reposed in the new generating authority were actually fulfilled, the aim of connecting practically every north of Scotland home to the national grid was triumphantly accomplished. Had it not been, the economic and other difficulties confronting the more rural parts of the region in the post-war years could well have become completely insurmountable.

The election of Labour to power in 1945, on the basis of a commitment to combine economic planning with a wholly new emphasis on social justice, ensured that the growing tendency for the state to become more and more embroiled in the affairs of the Highlands and Islands – something which had been observable since the 1880s and which had become still more marked since 1939 – would be maintained, if not accelerated. And this was particularly the case with regard to policy for agriculture.

The British farming industry had contributed mightily to the national war effort. But farmers had performed equally energetically in the course of the earlier conflict with the Kaiser's Germany only to see protectionist mechanisms – which wartime governments had been obliged to introduce in order to ensure that agriculturalists obtained an adequate return on their products – hastily scrapped in the early 1920s when low-cost food once more became available on world markets. The unavoidable outcome of that particular 'betrayal', as the farming lobby was inclined to call it, had been the price collapse of which crofters, along with practically everyone else in agriculture, had been the victims in the period prior to 1939. And it was by no means surprising, therefore, that the agricultural industry, again benefiting enormously from measures taken on its behalf at a time of national emergency, should have begun by regarding the incoming Labour government with a very wary eye. Would Clement Attlee, farmers wondered, do to them what David Lloyd George had done some 25 years previously?

They need not have worried. In 1946 the Labour Prime Minister – the first holder of his office ever to have attended this particular function – was the guest of honour at the annual dinner of the National Farmers Union of England and Wales. There Mr Attlee thanked his hosts for their endeavours, encouraged them to do still more and assured them of his administration's wholehearted backing. It was the beginning of an intimate relationship between government and the country's principal farming organisation which was to endure, largely irrespective of the party in power, until the 1980s.

Nor was this Labour-NFU connection, which was soon to be equally evident in Scotland, as surprising then as it might seem in retrospect. Farmers, even those operating on a substantial scale, had not long emerged from the grim years of the 1920s and 1930s. If they were not grindingly poor, neither were they the conspicuously affluent people that many of them later became. To the Attlee government, in any case, the wealth or otherwise of the farming community was probably less important than the contribution which the agricultural industry might be persuaded to make to the overall reconstruction of the national economy.

Putting the United Kingdom back on a sound financial footing depended primarily on expanding industrial exports. But the other side of the country's trade account also demanded attention. Any reduction in the British import bill, particularly for those commodities which were bought and sold internationally in dollars, would help restore the much battered balance of payments to some sort of equilibrium. And so it was thought increasingly important to augment domestic food production and to lessen the nation's dependence on cereals and meat purchased from countries like the United States and Argentina.

On this occasion, then, there was to be no prompt peacetime return to free trade. Instead British agriculture was to be sheltered from the overseas competition to which it had been exposed since the repeal of the early nineteenth-century Corn Laws. Food from other parts of the world would still be allowed into the United Kingdom, the Labour government made clear. But British farm prices, in marked contrast to previous practice, would not now be permitted to fall as a result. Rather than being determined, as in the past, by the vagaries of supply and demand, prices would be fixed annually by the government in negotiation with farming interests. Those prices, it was intended, would be pitched at levels high enough to encourage greater domestic output. And if – as was likely and as, in the event, occurred continually – the rate agreed for a particular commodity turned out to be above the price actually being paid for that commodity in markets to which imports still had access, then the difference between the market price and the guaranteed price would be made up by 'deficiency payments' financed from general taxation.

Nor was this the limit of the Attlee administration's generosity. Security of tenure, of the type long enjoyed by crofters, was now extended to tenant farmers also. Since the Hill Farming Act of 1946 was succeeded by the Agriculture Act of 1947 and the Agriculture (Scotland) Act of 1948, producers were provided with an almost bewildering array of financial aids designed to enable them to improve, expand and generally re-equip their farming businesses.

'The Government,' as the Department of Agriculture for Scotland noted in a summary of those wholly unprecedented developments, 'has decided in the national interest to stabilise the position of the primary producer by guaranteeing prices and assuring markets for the main agricultural products. The tenant farmer is further assured of continuity of tenure. In return for this stability and security the industry is asked for the greatest efficiency in production possible.'[4]

Just as crofters had profited from the livestock subsidies introduced in 1940 and 1941, so they gained, to some extent at least, from the still more generous state aid which became available as a result of the post-war policy for British agriculture. To the annual headage payments for which a crofter's sheep and cows already qualified, there was now added a subsidy for calves. Drainage grants became available; as did water supply grants, bracken control grants and an agricultural lime subsidy. A grassland fertiliser scheme ushered in a further set of supports. And a marginal production scheme was initiated to provide agriculturalists in more climatically disadvantaged localities, such as the Highlands and Islands, with cash incentives to reseed hill pastures, construct cattle shelters and install silage pits.

These measures could not fail to make an impact. In Shetland, for example, they helped to restore to production a good deal of the arable land which had gone out of tillage in the 1920s and 1930s. On much of that land, it was afterwards recalled, the growing of crops had simply been abandoned. And prior to 1939, of course, 'prices were not such as to tempt farmers to bring it back'. Things were very different in the years following the war, however. 'Prices are now assured,' it was reported from Shetland in 1951, 'and the very unfavourable farming conditions of the far north are recognised to some extent by subsidies.' In such circumstances, it had once again become a viable proposition to add to the area under cultivation.[5]

Nor did such land reclamation any longer involve the back-breaking toil with which it had necessarily been associated in the past. In the Highlands and Islands, as elsewhere in Britain, agriculture was being steadily mechanised. In 1942 there had been 16 caterpillar tractors in the whole of northern Scotland. By 1950 there were 149. The number of wheeled tractors rose from around 1,000 to just under 3,000 in the same period. And there was a still more dramatic increase in the region's stock of tractor-drawn ploughs and other implements.

But such statistics, though impressive, were not indicative of any marked amelioration in crofting conditions. Rather they reflected the extent to which farmers, as opposed to crofters, in those parts of the Highlands and Islands where the land was comparatively good and where holdings were generally bigger – places like Kintyre, the Black Isle, Caithness, Orkney and even, to some extent, Shetland – were able to take full advantage, unlike their smallholding neighbours, of policies geared primarily to the requirements of the country's larger agricultural producers.

Crofters, it should be stressed, were in no way legislatively debarred from access to the lavish financial assistance now on offer to the wider agricultural community. Both the Hill Farming Act of 1946 and the Livestock Rearing Act of 1951, for instance, authorised the Department of Agriculture to make grants for the improvement or rehabilitation of hill grazings of the type that were to be found throughout the crofting areas. And crofters were, in principle, fully eligible for such grants. In order to obtain them, however, they had to comply with a strictly stipulated set of conditions, which in the circumstances prevailing in all too many crofting communities, were simply impossible to meet.

Before approving a particular improvement scheme, the Department of Agriculture, as the grant-awarding agency, had to be satisfied that the scheme in question would create, or would make a worthwhile contribution to creating, a financially viable agricultural enterprise. As the Department's own officials explained, therefore, the first point to be considered with regard to any grant application was 'whether the holding for which improvements are proposed is big enough to stand on its own feet economically after rehabilitation'. As far as the Highlands and Islands were concerned, this implied, in the opinion of the relevant civil servants, that land improvement aid should be confined to holdings capable of carrying at least 300 sheep. Only a handful of crofts even began to approach such a size.[6]

Since farmers and crofters were not differentiated as such in the annual returns made by the Department of Agriculture in the 1940s and 1950s, it is impossible to be precise about the extent of the disparity between the two groups. But comparisons between different geographical areas can be made and these are revealing enough. They show, for example, that of the £450,000 being expended annually by the Department on Highlands and Islands land improvement projects in the early 1950s, only some £8,500 was going to Skye, a predominantly crofting locality, while Lewis, with its several thousand crofts, was getting barely £2,000 – less than half of one per cent of the total.

As these figures demonstrate, there was to be no crofting equivalent of the post-war renaissance in British farming. A visitor to Lochaber at harvest time in the later 1940s might 'come on the active township of

Bohuntin, tucked away in Glen Roy, and find there acres of golden oats on a southward-facing hillside'. But such discoveries were exceptional. Much more typical were the experiences of those members of one of Tom Johnston's many enquiry committees who 'were very much struck by the poverty-stricken appearance' of the typical crofting locality.[7]

'In many cases,' the committee recorded, 'the crofts are occupied by aged or infirm persons; in others they may have been bequeathed to a relative who is working in a city or may even be abroad, the result being that the holding is neglected and such arable land as may once have formed part of the subject has been allowed to revert to grass and rushes.'

Like their counterparts in the rest of the country, of course, the 'aged and infirm' crofters mentioned in that report were among the most obvious beneficiaries of the welfare state which was put in place by Attlee and his ministers. Though Malcolm MacMillan was by no means bereft of political bias in such matters, the Western Isles MP was doing no more than stating an evident truth when he commented that, 'because of the social security system, nobody in the Highlands can be said to feel the cold, hard edge of poverty as sharply today as in the pre-war years'.[8]

Nor were enhanced pensions, improved unemployment benefit and the inauguration of the National Health Service the only changes for the better occurring in the crofting areas. By 1951, for example, one in six crofts had already been connected to the Hydro Board's electricity supply system. Within another 10 or 15 years, that proportion would rise to four in five. 'There is no more heartening a sight in the Highlands,' it was understandably observed at that time, 'than to come on a distant crofting township and see the wires carrying power and light even to remote holdings maintaining a precarious existence on the edge of the wild.'[9]

In numerous other ways, too, the isolation of the average crofting community continued to be eroded. Inverness-shire County Council took the lead in attempting to establish youth clubs in the islands. Branches of the Scottish Women's Rural Institute were formed in a number of crofting localities. Village halls were built in many districts. And not the least popular events in many of these new community centres were the weekly or fortnightly 'picture shows' staged by the Highlands and Islands Film Guild which was launched in 1947 in order to bring the products of Hollywood and Ealing studios to even the most inaccessible corners of the crofting counties.

But none of these developments, however welcome, could disguise the brutal fact that the continued existence of more and more crofting townships was being put at risk by their all too apparent failure to retain the people needed to make them socially viable communities.

At the outbreak of war in 1939, as noted in the previous chapter, much smaller numbers of soldiers were forthcoming from the crofting areas than had been the case in 1914. That was because there were far fewer men for the conscription authorities to call upon. And when in 1951 the country organised its first census for 20 years, the depopulation of the Highlands was revealed to be proceeding as rapidly as ever. Practically all the crofting areas had lost a substantial proportion of their former residents. And in the worst affected localities, such as those parishes occupying the western seaboard between Cape Wrath and Applecross, the population had actually fallen by some 25 per cent since 1931.

'A decrease of population, however small, is usually a matter for concern,' observed the authors of one contemporary study of the depopulation issue. But when the loss amounted to 'a quarter or a fifth in the short space of 20 years', as it did in much of the West Highlands, then the situation could legitimately be reckoned 'grave indeed'.[10]

The township of Clashnessie in Assynt 'could be taken as an example of what was happening in the area', remarked one Sutherland crofter, Kenneth MacKenzie, in 1952. 'In 1910,' said MacKenzie, 'there were 138 people in Clashnessie; now there are only 20. The youngest "boy" is 28 and the youngest "girl" will not see 45 again.'[11]

Such distortions of the natural demographic order were becoming commonplace right across the Highlands and Islands. 'There were very few young people among the crofters,' it was reported from Ardnamurchan, for example. 'When they left school the young people went to work in the south and only came back for holidays.'[12]

Other districts experienced similar outflows. 'In many areas the proportion of elderly people is abnormally high,' noted one official analysis of crofting prospects. 'A tragic aspect of the decline of population,' recorded another survey, 'has been the great diminution in the proportion of children.' In such circumstances it was difficult to dissent from the singularly gloomy verdict of a paper prepared by a number of

Scottish Office civil servants in the summer of 1948. 'In brief,' they concluded, 'the Highland scene presents a picture of a drift of the younger people to the towns, leaving behind a diminished and ageing population which, in many of the remoter sectors of the area, is approaching a position at which it can no longer maintain itself.'[13]

Here, then, as was clearly recognised by one man who had spent a good deal of his time promoting the adoption of new farming techniques in the Highlands and Islands, was an obvious cause of the decline in crofting agriculture. 'There can be no sudden revival of crofting,' wrote Frank Fraser Darling, 'with the number and proportion of young people diminishing, and especially when the old methods of husbandry are being given up. An elderly man can do as much as a young one with the *cas chrom* or the spade or the scythe, but the older men are not tractor drivers, and where the young men have to go away from the townships to work, the husbandry of the crops is failing.'[14]

In the past, when there had been no alternative available to them, crofting communities had assiduously worked their meagre scraps of arable land with the help of simple implements like the *cas chrom*, the traditional foot plough of the Gael. 'Reliance on the croft for the bulk of subsistence demanded a relatively high standard of husbandry,' commented Fraser Darling, reflecting on the contrasts between that earlier period and the 1940s, 'and because a way of life consistent with subsistence agriculture was followed, it was possible to give the ground that standard of husbandry.'[15]

But that way of life had been fast disappearing even in the 1920s and 1930s. And by the 1940s it had vanished more or less completely. Nor, as indicated in this book's opening pages, was the demise of the former crofting order necessarily to be regretted. It was because the crofting population now had recourse to other – arguably much less oppressive – means of making ends meet, after all, that the Department of Agriculture was able to report in 1951 that 'the time is past when a croft, however small in size or poor in quality, is eagerly sought after as a means of partial escape from a state of destitution.'[16]

As Frank Fraser Darling observed of those developments: 'One cannot escape the conclusion that occupations other than working the crofts to their potential are sought first; the life of toil on the croft is second best.' And so barns and byres increasingly deteriorated, field drains went uncleaned, fences and dykes fell into disrepair, cattle numbers declined and many crofts were, in the end, simply abandoned by families who had moved elsewhere to live.[17]

When holdings thus fell vacant on a crofting estate, the owner of the land in question was required by law to notify the Department of Agriculture. Although the landlord, if he failed to find other tenants for such crofts, was legally permitted to retain them in his own occupation, he was not allowed to let croft land to people who were not crofters.

By the early 1950s, however, these obligations were being widely disregarded by Highlands and Islands estate managements. Many vacancies were simply not drawn to the attention of the authorities. Holding after holding was quietly abstracted from crofting tenure – a practice much criticised, naturally enough, by those crofters who still wanted to make something of the land.

'We have received many and varied complaints as to the manner in which some landlords dispose of vacant crofts,' reported the enquiry commission which eventually investigated these matters:

> Sometimes it is represented that the landlord refuses to let the holding at all, although – or so it is averred – a suitable tenant is available or the land could be used to enlarge adjoining holdings. At other times it is alleged that, instead of letting the ground to a deserving applicant, the landlord sells the croft over his head, without giving him a chance to purchase, or uses it to augment the holding of another crofter who has already several crofts in his hand, none of which may be properly worked. Again, it is complained that the croft is let or sold to a city dweller as a holiday house, instead of being granted to a *bona fide* crofter who would reside in the township and work his place as an agricultural subject. We recognise that in many of these cases the landlord concerned may have a good answer to these complaints made against him and that he, as well as the crofter, is the victim of powerful economic forces. Nevertheless, the complaints we have received are numerous and widespread and we are satisfied, on the evidence, that many of them are well founded.[18]

In those gradually disintegrating communities which were now so sadly characteristic of the crofting areas, especially on the West Highland mainland, it was difficult to be anything but dispirited. Shops, schools and churches were closing. Social activities of every kind were being given up. The Gaelic language was being spoken less and less. The consequent sense of things somehow falling apart was heightened by the presence, in so many townships, of large numbers of homes that were no longer occupied.

On at least 50 of the 330 crofts in Assynt in the early 1950s, the croft house was permanently uninhabited. Other homes were very seldom lived in, and the resulting spectacle of former dwellings steadily turning into ruins was both unsettling and demoralising. 'More than one person,' said one local government official with extensive experience of the crofting areas, 'had commented to him how depressing it was to see ruined houses and empty homes.'[19]

Here and there, most notably in the Western Isles, where depopulation was not quite so acute as elsewhere, traditional crofting practices were kept up. Cattle were still being driven to summer sheilings on the Lewis moorlands in the 1940s – as they had been driven by the women and children of the clans several centuries before. On the eastern side of Harris, in particular, crofters were still raising worthwhile crops on lazybeds or *feannagan* – those artificially created mounds of earth which the victims of the clearances had devised to provide themselves with a modest measure of cultivable land in the marshy and stony localities to which they were expelled by their landlords.

'Nothing can be more moving to the sensitive observer of Hebridean life,' thought Frank Fraser Darling, 'than those lazybeds of the Bays district of Harris. Some are no bigger than a dining-table, and possibly the same height from the rock, carefully built up with turfs and seaweed carried there in creels by the women and girls. One of those tiny lazybeds will yield a sheaf of oats or a bucket of potatoes, a harvest no man should despise.'[20]

The *cas chrom* remained in everyday use in places like Harris. And with its help, Frank Fraser Darling calculated, a man who was in reasonable physical trim could turn over a quarter of an acre in a 12-hour day. But for such toil to be anything other than unacceptably burdensome, it had to be undertaken communally. 'Digging by oneself is drudgery,' commented Fraser Darling, who had good cause to know, 'but shared with three others, with only elbow-room between them, it can be a joy.'[21]

In all too many crofting townships by the 1940s, however, it was practically impossible to put together such a work-team. Though agencies like the Scottish Agricultural Organisation Society – which had the job of promoting agricultural co-operation – successfully pioneered co-operative buying schemes and other helpful developments, it was extremely difficult for SAOS, or anyone else, to make a great deal of headway as long as crofting was so widely thought, by crofters themselves, to be in a state of virtually irreversible decline.

'The cultural integrity of the people has been invaded and their faith in the soundness of their way of living impaired,' concluded one analysis of the 1940s crofting scene.[22] Although they would not have been inclined to use such apocalyptic terms, SAOS fieldworkers were inclined to agree that the obstacles in the way of crofting revival were as much psychological as economic; which is not at all to say that SAOS was on the side of those who held crofters to be in difficulty simply because of their own lack of enterprise.

'Although it is often said that crofters are too conservative to change their ways,' SAOS reported in 1951, 'the Society does not believe this to be so. This apparent conservatism may, indeed, arise because there is undoubtedly a feeling of despondency amongst them and this will have to be overcome.'[23]

Given the necessary technical assistance, commented Roddy MacFarquhar, who worked with SAOS in the Hebrides in the years immediately following the war, those crofters still prepared to cultivate their holdings could make no end of agricultural progress. 'But the vital element required to bring the land into productive use, a young and vigorous population, has departed. This need to encourage young people to return to the land is the essential and primary problem that has to be solved.'[24] And as to why young people, contrary to Roddy MacFarquhar's earnest wishes, continued to leave most crofting townships, there were, as there had begun to be in the 1930s, no end of theories.

A generation less willing than its predecessors to accept the moral dictates of the various Presbyterian denominations which exercised such a key role in so many crofting localities, it was postulated, were rebelling by going off elsewhere. It was certainly the case that, whatever they believed they were accomplishing by campaigning against the construction of village halls or the formation of youth clubs, some ministers contributed to the extinction of the very communities they felt themselves called to serve.

'The church, they said, would not let them have a dance,' wrote the novelist Neil Gunn, reporting on a conversation he had in the early 1950s with a group of young women then preparing to leave a Lewis crofting district where, for many years, the only venue for a social gathering of that kind was the wooden deck of the bridge which carried the main road across a local river.[25]

But the Church was by no means the unrivalled choice of those in search of institutions and individuals to blame for depopulation. The Labour Government's educational reforms – which resulted in children having to leave island and west coast homes at the age of 11 or 12 in order to complete their schooling in distant towns – were said, for example, to have engendered a profound and widespread disenchantment with all things rural. And it was indeed difficult to find much evidence that the educational authorities saw it as their task to equip school pupils for life in a crofting area.

'He blamed the educational system for the diversion of attention from agriculture,' it was reported of Yell crofter, John Rendall, in 1952. 'Work on a croft was now regarded as derogatory.' Frank Fraser Darling, from his different perspective, was equally forthright. 'We did not have the situation in the Highlands,' he said, 'in which a fellow with brains could think he had a chance of using them in his home area.'[26]

Frank Fraser Darling's first wife, Marion, who worked extensively with young people in the crofting areas, was in full agreement. 'Many headmasters had told her that if they got a brilliant boy they sent him away to Inverness or Dingwall at once. That almost meant that they sent him away from the Highlands.' Such was the esteem in which the academically successful were invariably held, Marion Fraser Darling continued, that those young men and women who did remain at home on their family crofts were often thought to have failed, in some way, to make the best of life. 'There was an awful feeling in many townships that it was only the duds who were left.'[27]

But if ministers and teachers did little to impede the exodus of people from the crofting areas, the underlying causes of migration from the Highlands and Islands were not to be found in either the churches or the schools. They were to be discovered, in Marion Fraser Darling's opinion – and in this she was surely right – in the all too evident 'lack of work', on the one hand, and the equally obvious 'lack of amenities', on the other.

Crofters in the 1940s and 1950s were a good deal better off materially than were crofters in the 1930s; just as crofters in the 1930s had been incomparably better situated than their nineteenth-century predecessors. In relation to what was happening in most other parts of the United Kingdom, however, the progress being made in the crofting areas continued to be much less impressive. By the end of the 1940s, in fact, it was clear that the age-old economic gulf between the north of Scotland and the rest of Britain was, if anything, tending to grow even wider.

In both the Western Isles and Shetland, for instance, the wartime construction boom had inevitably ended. Traditional industries, such as fishing, knitwear and tweed, proved wholly incapable, at that stage at any rate, of generating the employment needed to make good the consequent loss of jobs – let alone provide for the large number of returning servicemen.

In Shetland, for the first two or three years after the war, unemployment was kept at bay by government-organised public works programmes. But these were terminated, with predictable results, in the later 1940s. In September 1948 there were only 90 men registered as jobless in all of Shetland. A year later, however, the total stood at 480. By January 1950 it had reached 1,000.

In the Western Isles meanwhile, the position was still more serious. At the end of 1946, when the national unemployment rate – thanks to the Attlee administration's full-employment policy and the overall economic recovery then occurring – stood at around 2.5 per cent, the comparable figure for Lewis was calculated to be in the region of 40 per cent. Although the Scottish Office established a series of committees to investigate the problem, the government itself soon made matters even worse by imposing a heavy purchase tax on island tweed which the Treasury, in the face of repeated pleas for a more enlightened attitude, stubbornly insisted on treating as a luxury item.

In such circumstances, island crofters could support their families only by taking jobs far from home. Many went to sea in the manner of their fathers and grandfathers – to the extent that it was estimated that one third of all the men nominally resident in Barra, for example, were actually serving with the merchant navy. Others joined the squads of itinerant navvies then labouring on the dams and power stations being built by the Hydro Board. But there were many young men, not surprisingly, who simply moved both themselves and their families to more prosperous parts of Britain – or, for that matter, to the United States, Canada or Australia – rather than entertain the prospect of spending the greater part of their lives on oil tankers or in work camps in order to maintain their wives and children, whom they would seldom see, on crofts which were, in any case, all too frequently bereft of comforts of a kind that were beginning to be taken for granted elsewhere.

The typical township's paucity of modern convenience was often evident even in the approach to it. The road serving one crofting parish in Argyll was, in the words of a Church of Scotland report of 1946, 'in an appalling condition. It is much too narrow for the amount of traffic it has to carry and the rough nature of its surface makes speedy and comfortable travel an impossibility. It is riddled with potholes and there are long stretches where rushes are growing to a height of six or seven inches in the middle of the roadway.'[28]

At the end of that road, the authors of the same report observed, were crofting communities where accommodation of any kind was often difficult to come by and, if available at all, invariably inadequate. 'The present state of housing is extremely unsatisfactory,' the churchmen concluded. Many couples were living in single rooms 'and endeavouring to rear families in very cramped conditions'. Many homes were damp, and many more had no 'sanitary arrangements'.

As late as 1947 it was calculated that about 40 per cent of all the croft houses in Lewis were still of the blackhouse type – with thatched roofs, rubble-packed stone walls and, very often, earthen floors. Although Department of Agriculture grants – in addition to the more long-standing loans – for the construction of croft houses were first made available in 1948, and were eagerly taken up, many of the new homes which resulted still had no proper water supply. Piped water was 'exceptional' in crofting townships even in the 1950s.[29]

'She had brought up a family in a croft house with no sanitation,' noted the members of one enquiry commission to which Marion Fraser Darling was invited to give evidence. 'If asked whether she would rather have electricity or water and sanitation, she would say you can keep your electricity if you give me water and sanitation. Electricity was good, but not nearly so important as water.'[30]

She knew townships where there were 'practically no girls', Marion Fraser Darling went on. 'Again and again you would find 15 boys or young men in a crofting township, but to get one or two girls was unusual.' Something had to be done 'to keep the girls there'. And, in Marion Fraser Darling's opinion, the provision of adequate water supplies was crucial. The modern woman, she said bluntly, was not prepared to make a home in a place where it was impossible to install a kitchen sink, let alone a flushing toilet or a bathroom.

The commission to which Marion Fraser Darling made her comments was, it seemed, fully persuaded by her arguments. 'If it is a question of comfort, convenience or entertainment,' ran the commission's report, 'crofting life compares unfavourably with urban conditions, especially for women. The crofter's wife has to take a hand in all the operations on the croft and, over and above the cares of house and family which she shares with her sister in the town, she may have to carry water from the well, often some distance away, and to bring the household supplies along township roads which in winter may be not only inconvenient but actually dangerous. It is no great exaggeration to say that the key to the whole crofting problem lies in the hands of the women, especially the young women. If they elect to stay in the township, there is hope for the future. If they leave, they will probably never return.'[31]

To any 'observant person' travelling through the crofting areas, commented the *Oban Times* in 1952, 'certain facts are all too obvious'. Of these, the paper continued, the 'two most deplorable' were 'the prevalence of buildings in ruinous condition' and the 'incursion of rushes and bracken on to the arable land of the croft'. These abuses, so the *Oban Times* leader-writer thought, could fairly generally be attributed to the spread of absentee tenancy.[32]

> In the great majority of cases, what happened was something like this. Some 20, 30 or 40 or more years ago the son who succeeded his father in the tenancy (having received what was then generally considered a 'better' education than his father) decided that the croft was too small to give him the standard of livelihood he desired for himself and his family. He saw in it nothing but hard work and comparative penury for the rest of his days. So he decided he would try his luck elsewhere.
>
> But when he decided to go elsewhere for a livelihood, rarely did he renounce the tenancy of the croft. For the desire to retain a claim on the wee place that for generations was the home of his forebears is one of the strongest elements in a crofter's make-up. So he would sublet to a neighbour.
>
> The fact that, under the Crofters Act, in absence of the landlord's

consent, a sublet could be considered null and void did not trouble him at all; he never asked the landlord's consent. He just sublet to his neighbour and went off to Glasgow or Canada or America. And whether to his credit or otherwise, it must be said that neither did the landlord bother about the matter. Sometimes he got rent direct from the illegal subtenant, sometimes from the absentee whose name was continued in the rent-roll as tenant of the croft; and always the receipt was made out in the name of the latter or his representative.

Meantime the subtenant needed only one house and steading. Should those on his own croft be the better conditioned, he would continue to use them and let the others go to ruin. Should those on the sublet croft be the better, he would occupy them. In either case, one set of buildings was sure to go into decay.

Then with regard to the land, the natural tendency was to put as little into it, and take as much out of it, as possible and apply the resulting manure to his own croft. Nor, in the case of the sublet land, is there likely to be great concern to keep fences and drains in good order. Hence these sad blots of ruined houses and neglected land.

Absentee tenancy of the type described by the *Oban Times* appears to have come about, in a very real sense, by accident. It did not exist in the years following the passing of the Crofters Act of 1886 – the security of tenure introduced by that act being enjoyed, reasonably enough, only by those crofters who resided on their crofts. And it became possible only in 1917 when the Court of Session, adjudicating on a test case, interpreted the somewhat casually drafted wording of the Small Landholders (Scotland) Act of 1911 – a measure which, to some extent, superseded the 1886 legislation – in such a way as to result in security of tenure no longer being conditional on the permanent occupancy of the holding to which that security of tenure applied.

Prior to 1917, then, a crofter who moved elsewhere was obliged to renounce his tenancy – and his landlord was equally obliged to make his vacant holding available to a new occupant. After 1917, however, any crofter who chose to start a new life in another part of the world was free to retain his tenancy. Many migrants did so; often, no doubt, in the hope – only occasionally realised – of one day returning.

For the absentee to sublet his holding without first obtaining the written consent of his landlord was, as the *Oban Times* pointed out, a breach of crofting law. But it clearly suited the absentee to have his rent paid by a subtenant – usually a neighbouring crofter who typically allowed the absentee's fields to degenerate into rough pasture. It was equally convenient for the proprietor to leave well alone. Had the absentee chosen to renounce his tenancy in the pre-1917 manner, after all, the landlord would have been required to compensate him with a sum equivalent to the value of any permanent improvements, including the croft house, which his holding might contain. As depopulation gathered pace, there was no guarantee of estate managements being able to recoup such outlays from incoming tenants. All too frequently, there were no tenants to be got.

'Cases have been cited to us,' one enquiry committee reported, 'of crofters, entitled as things are to all the rights conferred by the Crofters Acts, including security of tenure, who are permanently resident in the United States while the croft house remains closed from one year to another and a few sheep from a neighbour's holding graze over the untended fields.'[33]

In one township which the committee examined especially closely there were 13 crofts, all but one of them equipped with a reasonable house. 'Seven of the crofts are occupied by absentee tenants whose homes are in Glasgow, Motherwell, the United States, Edinburgh, Conon Bridge, Greenock and Australia,' the committee commented. 'They use the houses only for holiday purposes.' This was an 'extreme case', the committee acknowledged. 'But the same kind of thing is happening in many places.'

Absentee tenancy, vacant crofts, derelict land, abandoned homes and massive depopulation, when taken together, added up to powerful evidence of a society in an advanced state of disintegration. In the face of all of this, however, the post-war Labour Government seemed curiously inactive, even indifferent. Certainly no attempt was made to resuscitate the concept, as advanced by the Hilleary Committee and others in the 1930s, of providing the Highlands and Islands with new developmental institutions.

This was something which Tom Johnston, a little surprisingly, had very firmly opposed. The Secretary of State had just returned from the Highlands, Ross and Cromarty county councillors were informed by the Scottish Office in June 1942. 'His discussions there have confirmed him in the view that it would be a fundamental mistake for the Highlands to draw attention to themselves as a depressed area in need of special treatment. Such a course would not be likely to attract new industries or facilitate post-war development; rather the reverse. It seems to the Secretary of State of supreme importance that Scotland as a whole should stand together as a single unit so as to secure her fair share of industrial development.'[34]

At the end of the war, some Scottish politicians, most notably John M. Bannerman of the Liberal Party, were still pressing for what Bannerman called a 'co-ordinating development authority' for the north of Scotland. Such an agency, Bannerman said in Glasgow in 1945, 'should have sufficient financial backing to carry through the comprehensive rehabilitation of the Highlands'.[35] That much had been common ground in 1939. But now it was Tom Johnston's much less forward policy which was to prevail – despite its orginator having decided to quit active politics in order to assume the chairmanship of his beloved Hydro Board.

Johnston, of course, was prone to regard his generating agency as constituting a development authority in its own right. Although this was an exaggerated view, it was not one likely to be challenged by Joseph Westwood, the 1945 Labour Government's Scottish Secretary and a man who has been described by one authoritative historian of that star-studded administration as 'perhaps the least impressive member of the Attlee Cabinet'.[36]

Both Westwood and Arthur Woodburn, appointed Secretary of State in 1947, were convinced centralisers who, if they gave any worthwhile thought to such issues, would have tended to assume complacently that the economic problems of the Highlands and Islands, just like those of Scotland generally, were not such as to require the implementation of measures additional to those being applied by the Labour Government to Britain as a whole. Nor was there, at that time, any very powerful body of contrary opinion.

Inside the Scottish Office, for example, Highlands and Islands issues were thought in 1945 to merit no very urgent attention – a situation in sharp contrast to that prevailing, some 25 years earlier, in the immediate aftermath of the previous world war. Then the Board of Agriculture for Scotland had been dealing very largely with crofting questions. Now these were a quite subordinate part of its successor organisation's day-to-day concerns.

Highland policy in the 1940s – as was to remain the case until the 1960s – was the virtually exclusive responsibility of the Department of Agriculture. But the post-war Department – which was having to handle the administrative consequences of the government's much more interventionist attitude to national food production – was a very different institution from its pre-war counterpart. Its staffing complement had soared from well below 700 in 1939 to more than 2,000 nine years later. If Frank Fraser Darling was overstating his case when he remarked in the early 1950s that the Department of Agriculture 'was not particularly interested in the Highlands', there is no doubt that the Department's senior officials were by then inclined to give a far lower priority than their predecessors to purely Highland matters.[37] Men who had been asked, in effect, to take over the management of the entire Scottish farming industry, after all, could scarcely be expected to worry greatly about crofting – the more so since crofters themselves were doing comparatively little to press their problems on the Scottish Office's attention.

This was the other great contrast with what had gone before. When, in 1919, thousands of demobilised servicemen returned to the crofting areas, they promptly set about the illegal occupation of the sheep farms from which their ancestors had been evicted – thereby forcing Lloyd George's coalition government to embark on a major programme of land settlement in order to satisfy the demand for crofts. But there were few such protests in 1945 or in the years that followed. Only on two occasions, in 1948 and 1952, was land seized in the previously common manner.

The first such episode occurred on the Knoydart estate which consisted of a remote, mountainous and largely roadless peninsula on the western seaboard of the mainland some miles north of Mallaig. Knoydart was owned by the singularly unpleasant Lord Brocket who, in the 1930s, had been a fervent admirer of Adolf Hitler – at whose invitation he had attended a number of Nazi Party rallies and other functions in Germany. So notorious was Brocket's mismanagement of his extensive Highland property that it was requisitioned by the wartime government in 1940 in order to ensure that a reasonable number of sheep were pastured on it.

Knoydart was subsequently restored to Brocket. But despite the fact that well over 2,000 sheep had been added to the estate flock when it was in public ownership, Knoydart's total sheep stock, by 1947, had fallen below even its 1940 level. Since this had occurred at a time when everything possible was being done to maximise Britain's agricultural output, consideration was being given to requisitioning the property once again.

At this point, however, a number of Knoydart men formally asked the government to undertake a settlement scheme on the estate where, as they pointed out, there was no lack of land of the sort needed to form new holdings. And in November 1948 – by which date it was obvious that, for all Brocket's manifest failings as an agriculturalist, the authorities were not anxious to make any part of his property over to crofters – the Knoydart men, who had previously confined themselves to petitioning the Secretary of State, took more forceful measures to enforce their demands. They occupied a number of Lord Brocket's fields on which they then staked out several 'crofts'.

This particular type of proceeding, when adopted in places like Skye, Raasay and the Western Isles in the 1920s, had usually resulted in official action being taken on behalf of the protesting crofters. But for all that the earlier land settlement legislation was still in place and could, in principle, have been easily implemented, no such action was forthcoming in the case of Knoydart. Indeed the Labour-controlled Scottish Office seemed every bit as determined as the estate's pro-Nazi proprietor not to give way to the demands of the land raiders whose campaign, as a result, ended in total failure.

Other protestors at Balmartin in North Uist were more successful – though their objective was not to have new crofts created but to have their existing holdings enlarged at the expense of an adjacent farm.

'We have great sympathy with many crofters confined at present to meagre holdings to which their ancestors were driven in the clearances,' commented the members of an enquiry commission in 1954, 'and which march with large farms on which the crofters cast envious eyes.' This was the precise position at Balmartin.[38]

They wished to draw the attention of the Department of Agriculture, remarked Balmartin's tenants in a letter written in the early part of 1951, 'to the very severe difficulties and hardships experienced by us as crofters . . . in our endeavour to wrest a living from our crofts'. Their township contained seven holdings of seven acres each on 'very poor, light, sandy soil'. Their hill grazings were equally unproductive. It was, in the Balmartin men's opinion, 'very unreasonable to expect any

71

crofter to make a living under such impossible circumstances'. Accordingly they had 'come to the conclusion that in order not to break up our homes, which we on no account wish to happen, there is no alternative open to us but to seek more land'.[39]

Such land, the Balmartin crofters felt, was readily available on the neighbouring farm of Balelone. 'This 850 acre farm is at present under the proprietorship of an English gentleman who, we understand, already owns large estates elsewhere. The mere loss of one small farm would in no way affect the livelihood of a gentleman of means.'

But the gentleman in question, Lieutenant-Colonel Henry John Cator of Woodbastwick Hill, Norwich, understandably enough thought otherwise. Balelone's owner, it was clear, was not prepared to relinquish any part of his property. The Department of Agriculture was equally unprepared to exercise its powers of compulsory acquisition. The Balmartin crofters, after much procrastination on the part of the authorities, were eventually informed that there was nothing that could be done for them. The Department of Agriculture doubtless hoped that this was the end of the matter. It was not to be so.

'We have very reluctantly decided to take action on our own,' the Balmartin crofters informed their MP, Malcolm MacMillan, towards the end of 1952. 'We are going to raid the farm of Balelone on the 28th day of November. By so doing we shall give the whole world an opportunity to judge the righteousness of our request and actions. We have nothing to hide. It is our intention to give ample warning to all parties concerned and to conduct ourselves in as orderly a manner as possible in the circumstances. This is no idle talk on our part. We are on no account going to turn back.'

Among the letter's signatories were Roderick MacDonald, Alexander MacVicar, Donald MacDonald and John MacKay. And at nine o'clock in the morning of 28 November 1952, as they had promised, these four men marched smartly down the track leading to Balelone Farm.

Finding the farm gate locked they climbed a wire fence and began pacing out their various claims – which they identified by driving in several wooden stakes and by carefully cutting their initials, with spades brought for that purpose, in the farm's spongy turf.

There had been scores of similar episodes in the Hebrides since the 1880s. Many had been violent – involving, on occasion, pitched battles between crofters and the police. This, the last land raid, was more peaceable, and it resulted in a modest crofting victory; for, after a good deal more negotiation and discussion, the Balmartin crofts were eventually enlarged as their occupants had requested.

Had the 1945 Labour Government been under more pressure from crofters, it might have been prepared to apply its agricultural policies rather more imaginatively in the Highlands and Islands. But the generation then coming to maturity in the crofting areas was not greatly interested in following the example set in Knoydart and North Uist. Their future, so thought most young people from a crofting background, lay elsewhere. The Department of Agriculture's senior civil servants – well aware that the north of Scotland was not contributing as much as it might have done to the national effort to boost food production – were increasingly inclined to treat the crofting population's departure as an opportunity to make land use patterns in the Highlands and Islands conform a little more closely to those prevailing in the agriculturally more successful parts of the country.

To one set of observers in Lewis, such attitudes demonstrated a lamentable 'lack of flexibility'. Centrally determined policies were being applied 'without any modification to meet local cirumstances'. The many distinctive features of the crofting condition were going unrecognised. The crofter, for all his range of interests and involvements, was 'regarded as a small farmer and treated as such'.[40]

The Department of Agriculture, however, was unmoved by such complaints. The Hilleary Committee's earlier emphasis on the need to ease crofting difficulties by expanding non-agricultural employment in the crofting areas was set aside entirely. The crofting problem, in Scottish Office eyes, was essentially an agricultural one – to which agricultural solutions ought ideally to be vigorously applied.

What was required, a Department of Agriculture report on the north of Scotland concluded in 1947, was 'the amalgamation of crofts into much larger units'. There was, of course, one drawback to such measures – as the report's authors reluctantly acknowledged. They might involve an unacceptable degree of coercion. 'This . . . could be the policy of evicting the crofters which was so unpopular in the nineteenth century and which is unlikely to find favour today.'[41]

Such an approach to crofting issues was not calculated to appeal, on the face of things, to Hector McNeil who became Secretary of State for Scotland in 1950, following Labour's narrow retention of power in that year's general election. Previously a Foreign Office minister who had spent a good deal of time with the fledgling United Nations Organisation in New York, McNeil was a more impressive politician than his immediate predecessors. Since his father's family came from Barra, and his mother's people from Islay, the new Scottish Secretary might have been expected to identify strongly with crofters.

But McNeil's principal Highland hero was no crofter. He was the American-born landowner, J. W. Hobbs, who had bought the Inverlochy estate at the southern end of the Great Glen in order to demonstrate that cattle rearing in the United States style was not at all impossible in a Highland setting.

The Inverlochy experiment caught McNeil's imagination. It seemed to him to offer a way of bringing about a dramatic revival of agriculture in the Scottish hills. He personally visited Inverlochy, and he went so far as to establish a committee to investigate means of more widely applying the techniques so strikingly pioneered by Hobbs.

But in the many tributes then being paid to the Lochaber 'rancher' in the Scottish press, one aspect of the Inverlochy developments was seldom commented upon. The entire venture depended on Inverlochy's proprietor buying out his crofting tenants in order to add their holdings to his own farming enterprise.

That Hobbs found it financially worthwhile to embark on such a policy was itself due to the fact that, thanks to the various hill cattle subsidies introduced by the Labour Government, it was beginning to make sense, from a landlord's point of view, to think about ways of utilising croft land for other – more profitable – purposes. A similar conjunction of financial circumstances had brought about the clearances. Though Hobbs was no Patrick Sellar, the obvious implication of his activities, and of the Scottish Secretary's evident admiration of them, was that the continuation of crofting was incompatible with the type of agricultural progress which the government wished to encourage in the Highlands and Islands.

This did not augur well, from a crofting standpoint at any rate, for the Commission of Enquiry into Crofting Conditions which Hector McNeil appointed in 1951. It is not particularly surprising, therefore, that more pro-crofter politicians, notably Malcolm MacMillan, should have viewed that commission's emergence with the deepest of suspicions.

Chapter Three

LOOKING FOR A NEW APPROACH

In the course of the various wrangles following the publication of the Hilleary Report in 1938, a Scottish Office civil servant had suggested that, if a Highlands and Islands Development Board or any other authority of that type were to be ruled out on grounds of cost, 'consideration might be given to the appointment by the Secretary of State of a Highlands and Islands Advisory Council.'[1] And this, as it turned out, was the course of action eventually favoured by the stolid Joseph Westwood to whom such an apparently innocuous body was infinitely preferable to the more powerful agency demanded by the likes of John M. Bannerman.

The formation of an Advisory Panel on the Highlands and Islands was accordingly announced in the House of Commons by the Scottish Secretary in December 1946. Its members, Westwood made clear, would have no executive responsibilities. But the various parliamentarians, local councillors and other notables appointed to the panel, which was to be chaired by Malcolm MacMillan MP, would be free to examine such issues as they saw fit. And among the matters in which the Secretary of State's newly installed advisers were to take an interest, it gradually emerged, was the state of crofting.

Particularly anxious to propel the panel in this direction was the Department of Agriculture – its staff becoming more and more agitated about the implications of the fact that, at a time when national policy was geared so ostentatiously to obtaining substantial increases in food production, and when such increases were actually being recorded in most parts of the United Kingdom, the agricultural output trends in the Highlands and Islands were practically all downwards.

There was 'scope for increasing the agricultural production of the Highlands in the form of livestock and livestock products,' the Department insisted in June 1948, adding pointedly that the products in question were those 'which are the most expensive to import and which at present are in shortest supply'.[2]

Why, then, was the north of Scotland not performing better? One obvious answer, in the judgment of the Department, and in the opinion, too, of those other interests which now began to ally themselves with its mounting campaign for something drastic to be done about crofting, was to be found in the extent to which serious obstacles to agricultural production were inherent in the nature of crofting tenure.

The security of tenure principle, as applied to crofting in 1886, was no bad thing in itself, the Department conceded. Such security, after all, was extended to tenant farmers by the Agriculture (Scotland) Act of 1948 – a measure which the Department had very largely framed. But that Act, Department officials observed pointedly, made provision 'for the supervision and direction of owners and occupiers on grounds of bad estate management or bad husbandry with powers to dispossess in the event of failure to improve'. And it was high time, in the Department's opinion, that crofters, as well as farmers, were subjected to such disciplines.[3]

In 1947 the Department of Agriculture had asked Highland Panel members for their reaction to the proposition that crofting tenancies might be terminated in cases where crofts – such as those held by absentees – were not being properly worked. And though the panel was less than anxious to become embroiled in such a delicate matter at that stage, its members simply refusing to answer the question put by the Department, the crofting issue was clearly going to have to be confronted eventually.

Soon the Scottish Landowners Federation, the organisation representing the principal crofting proprietors, was weighing in with the thought that, 'if food production was to be the ruling principle, the question of amalgamating crofts . . . would have to be considered'. The County Agricultural Executive Committees – set up to implement the 1948 Agriculture Act – were complaining that 'the standard of husbandry in the crofting districts' was 'low' and calling, like the landlords, for a programme of croft amalgamation. The Agricultural Colleges, too, believed strongly that crofting had had its day. Even the Scottish Agricultural Organisation Society, which could generally be trusted to put in a good word for crofters, was clearly losing patience with its

Highlands and Islands clientèle: [4]

> As regards security of tenure, the interpretation of the Crofting Acts has
> given crofters a greater degree of security than that enjoyed by any other
> class of farmer in Scotland. It is exceedingly doubtful, however, whether
> crofters have made use of their advantages in this respect. Instead of more
> or less absolute security of tenure having encouraged them to improve their
> holdings, as one would expect, their immunity from fear of eviction some-
> times appeared to the observer to have engendered among them a degree
> of complacency and a disregard of the need to move with the times which
> has resulted in many crofting townships too often being among the most
> backward of farming communities. [5]

It was against this background that the Highland Panel, in the
autumn of 1950, finally began to get to grips with crofting problems.
'The question whether the Landholder Acts were now a clog on
Highland development must . . . be answered,' said one panel member.
'It would be of great value if crofting townships could be reorganised,'
remarked another. Soon the panel's secretary was recording 'a general
feeling that the best use was not being made of croft land' – and noting
an equally strong opinion to the effect that steps would have to be taken
'to get rid of absentee tenants'. But there was also, on the part of panel
members, a shrewd appreciation that 'the problem needed careful
handling' and that nothing should be done precipitately. [6]

It was at this stage in the debate, which was still being conducted very
much behind the scenes, that it began to be recalled that an important
preliminary to the Crofters Act of 1886 had been the Royal Commission of Enquiry chaired by Lord Napier. Perhaps the Secretary of
State ought to be sounded out on the possibility of initiating another
such investigation as a prelude to a new round of crofting reforms?

Such an approach was particularly attractive to Highland Panel
members, one civil servant cynically suggested, because it would relieve
them of any responsibilities in the matter: 'While the panel have
ventilated this controversial problem . . . they are by no means anxious
to accept the onus of suggesting a solution.' [7]

But the Scottish Secretary, Hector McNeil, was nevertheless
prepared to be amenable. It would indeed be helpful, he made it known
to Highland Panel members in November 1950, if they were formally
to advocate the establishment of a Commission of Enquiry. 'It would

be even better,' McNeil added in a note to one of his senior officials, 'if the Panel indicated to me that they were dissatisfied about the size of crofts and the level of agricultural enterprise on these crofts'.[8]

The panel duly obliged. Its members had recently 'been giving special attention to the condition of the crofting counties', it was announced publicly at the conclusion of a Highland Panel meeting in Edinburgh in December. And though they remained of the view 'that the future of the greater part of the Highlands and Islands depends in large measure on the retention of an active smallholding population making full use of the agricultural resources of the area,' they felt 'bound to draw attention to the fact' that 'these conditions' did 'not exist' in much of the north of Scotland.[9]

The Highland Panel statement continued:

> They recognise that there are areas which are substantially holding their population and where agricultural production is being maintained or even expanded, but there are also areas where population is still declining and agricultural production falling. Even where the signs of decline may be less obvious, there are vacant or uncultivated crofts, absentee tenants, inadequately sized holdings . . . . There are also barriers in the way of land improvement by enterprising crofters . . . . The Panel feel, therefore, that the time has come when serious attention must be directed to the difficult questions involved in strengthening a community largely dependent on the soil.

What was required, panel members concluded, was that 'the whole subject should be reviewed in the same authoritative manner in which it was reviewed by the Napier Commission in 1884'. Accordingly, they were making this opinion known to the Secretary of State.

Not the least striking aspect of the Highland Panel statement was the extent to which it faithfully reflected the now prevalent Scottish Office belief that the crofting problem was, in essence, an agricultural one – arising from causes such as the prevalence of diminutive holdings, aggravated by phenomena such as crofter absenteeism and resulting, or so it was increasingly alleged, in the failure of the crofting localities to contribute properly to the national food production effort.

Highland Panel members, to be fair, were not unaware of the fact that the typical crofter had traditionally depended on non-agricultural activities for much of his total income. Indeed their public statement of December 1950 had made passing mention of there 'being insufficient sources of supplementary employment' in many crofting areas. But no stress was now being placed – as it had been so conspicuously by the Hilleary Committee more than 10 years earlier – on the need to reinforce the crofting structure by means of diversifying the overall Highlands and Islands economy. On the contrary, it was beginning to be accepted, as the Department of Agriculture so clearly wanted it to be, that the way to deal with crofting was to make it agriculturally more efficient. And that implied, quite evidently, that the tenurial arrangements underwritten by the Crofters Act of 1886 would have to be altered fundamentally in order to produce the larger holdings which the Department thought to be so necessary.

Malcolm MacMillan, who had been absent from several of the Highland Panel's key sessions on the crofting issue, urged the Secretary of State 'to exercise caution and second thoughts before plunging into an enquiry'. The prominent Church of Scotland clergyman, Revd Tom Murchison, who came from Glenelg and who took a keen interest in Highlands and Islands policy, wondered publicly about the wisdom of interfering with the crofting community's hard-won legal rights. But this was very much the minority view.[10]

Much more representative were newspapers like the *Glasgow Herald* and *The Scotsman* which reacted to the Highland Panel's call for an enquiry by urging 'the consolidation of small crofts into more productive units' and by condemning those crofting laws which permitted a Highland landholder to have security of tenure 'even though he lives in Australia'.[11]

Some months later, at a meeting held in Stornoway to consider how Lewis crofters should respond to these developments, one man was seen to be carrying a home-made placard. 'Hands Off the Crofters Act', it read.[12] But it was by then a certainty that crofting legislation was about to be reformed.

The much canvassed Commission of Enquiry was finally established by Hector McNeil in June 1951. Its remit was 'to review crofting conditions in the Highlands and Islands with special reference to the secure establishment of a smallholding population making full use of agricultural resources and deriving the maximum benefit therefrom.' The Commission's chairman, the Scottish Secretary announced, would be

Thomas Taylor, Principal of Aberdeen University and a prominent lawyer and educationalist who leaned politically towards Labour.

Among the several other members of the Commission were the Glasgow University economist Alex Cairncross, the leading Scottish literary figure Neil Gunn, the prominent Ross-shire farmer Alasdair MacKenzie and two crofters, Robert MacLeod from Lewis and Margaret MacPherson from Skye.

Taylor and his colleagues first met on 29 June in Inverness. There they decided to issue 'a general invitation to organisations and individuals to submit evidence'. They also agreed 'not to hold their meetings in public, the general feeling being that certain witnesses might feel handicapped in giving evidence with members of the public and the press present.'[13]

This was the first of a number of differences between the Taylor Commission and the Napier Commission of some 70 years before. The earlier body had held most of its sessions in public and had relied very heavily on the personal testimony of crofters – that testimony accounting for the bulk of the material to be found in the four hefty volumes of evidence which the Commission ordered to be printed.

The Taylor Commission, though its members travelled to the crofting areas and though they held a number of meetings with crofters, relied rather more on evidence provided by organisations such as the Department of Agriculture, the various Agricultural Executive Committees, the Forestry Commission, SAOS, Highlands and Islands county councils, the Hydro Board and the Scottish Council of Social Service. Lengthy submissions from official or semi-official sources of that sort fill many of the files now reposing on those Scottish Record Office shelves which are devoted to the mass of paper generated by the Taylor enquiry. The opinions of crofters, in contrast, occupy no more than a few type-script pages.

That this is so was not the Taylor Commission's fault. Napier had been conducting his investigations at a time when much of the north of Scotland was in a state of virtually open rebellion and when crofters were engaged in highly aggressive political action designed to secure a well-rehearsed set of reforms. But in the 1950s, as opposed to the 1880s, the senior representatives of the bureaucracies dealing with crofting administration seemed to be more active politically – and certainly carried a lot more weight with government ministers – than crofters themselves.

It was not simply that the latter lacked an organisation capable of doing what the Highland Land League had once done so effectively. The crofting population as a whole, if not totally demoralised, was by no means unanimously convinced that the crofting life could ever be made sufficiently attractive to prevent the departure of more and more young people from crofting areas. While it was certainly the case that the perennially outspoken Margaret MacPherson, for one, was not likely to let her Commission of Enquiry colleagues forget that crofters were entitled to some say in their own future, there was no very obvious consensus – as there had so clearly been in the 1880s – as to what that future ought to be.

The crofting community, in other words, was extremely vulnerable to external pressures of the type being generated so assiduously by the Department of Agriculture. In such circumstances, there seemed good reason to believe – as Malcolm MacMillan and a number of his Western Isles constituents certainly believed – that any changes recommended by Thomas Taylor and his colleagues were very likely to be changes for the worse.

'In the course of our travels,' the Taylor Commission's members were to comment in their published report, 'we have been impressed with the sense of expectancy with which, in many places, our visit has been awaited.' [14] The records of the Commission's actual proceedings, however, tell a slightly different story.

There were, beyond doubt, a number of localities, such as parts of Shetland, where the Commission was very warmly received and where its members found a 'widespread demand' for some relaxation of the legislative controls which were so generally considered, in official circles, to be impeding agricultural expansion. But there were other places, most notably in the Western Isles, where the Commission's reception was anything but cordial. [15]

'The Commission's visit was not greeted with any great enthusiasm in Lewis,' the Commission's secretary noted tersely. 'At some centres very few people came forward to give evidence. At Carloway, where there was a big turn-out, this appeared to be due to a certain suspicion of the purpose of the enquiry. The contrast with Shetland was quite striking. In Lewis there was no sense of expectancy that material benefits might come from the enquiry. Rather there was an underlying fear that the Commission's object might be to undermine the Crofting Acts.' [16]

Nor were the distinctions between Shetland and Lewis the only ones the Commission was going to have to reconcile. Sometimes indeed it seemed that there were no two sets of people with identical views as to how crofting should develop.

College of Agriculture economists told the Commission that the future of the Highlands and Islands necessarily lay in 'hill cattle and hill sheep farms', or, failing that, in a 'forest economy'. The Department of Agriculture was scarcely more constructive. The Agricultural Executive Committees, composed almost exclusively of larger farmers, were dismissive of the crofting community's agricultural abilities and offered few positive suggestions other than the now standard comment that ways would have to be found of enlarging crofts.[17]

Landlords called for the dispossession of absentee tenants. Some tenants, for their part, were equally anxious to be rid of absentee landlords. In parts of Sutherland, it was alleged, proprietors of that sort were simply refusing to let their crofts. 'The landlord,' said one man in Lochinver, 'would like to see the end of the last crofter in Assynt.'[18]

By Frank Fraser Darling the Commission was advised to do nothing that would alter the underlying fundamentals of crofting tenure. The Crofters Act of 1886, said Fraser Darling, had been 'absolutely necessary'. The Commission would be 'very foolish' to tamper with it. 'The 1886 Act was very real in the mind of every crofter.'[19]

Others, however, were equally convinced that the principles of 1886 should, by one means or another, be wholly overturned. Those pressing for wholesale amalgamations were clearly in this category. But so were the one or two individuals who raised their voices in favour of converting crofting tenants into owner-occupiers – an idea of which much more was to be heard in the 1960s. And so, too, were the advocates, the most forceful of whom were to be found in the Labour Party's Shetland branch, of nationalising all croft land.

There was much, then, for Commission members to consider. And there was, as one of their early working papers put it, 'an important decision on policy' to be taken at the start: 'namely, whether the existing rights of crofting are to be regarded as sacrosanct, even when they conflict with desirable development; or whether the existing rights of crofters must be made to yield to wider considerations of general public interest'.[20]

At the centre of the clash thus implied between a 'public' and a 'crofting' interest, of course, was the constantly recurring issue of whether crofts should be amalgamated in order to enable croft land to contribute more to national food production. And if such was, in the end, to be the Commission's overriding objective, it might even make sense 'to abolish crofting legislation as a separate code and to bring the crofter under the general agricultural legislative system.'

The Department of Agriculture might well have welcomed such a recommendation. But the Taylor Commission did not go quite so far: 'While this would be perfectly practicable, and tidier, the wholesale repeal of the separate crofting legislation would probably be misunderstood by the public and be consequently unpopular.'

This, perhaps, was to put the matter mildly. It was also to dodge the prior question of what the Commission's central purpose ought to be. And so the chairman sought to bring the discussion back to basics. 'It may be asked,' Thomas Taylor commented in the course of a memorandum which he compiled in April 1953, 'whether the time has not come to supersede the landlord in the crofting counties.' Such a course had been 'urged' upon the Commission by a number of witnesses. Not least because of the extent to which the powers of crofting landlords had been circumscribed by earlier legislation, Taylor recognised 'the logical force' of moving in that direction. He was 'doubtful, however, whether the terms of our remit entitle us to consider this solution'.[21]

Margaret MacPherson harboured no such reservations. 'The Commission would be taking a big step if they recommended any modification of security of tenure,' she told her colleagues in the early part of 1953. Some such modification might be necessary in order to deal with crofter absentees. But if that 'had to be done', Mrs MacPherson contended, 'it would only be fair to come down heavily also on the useless landlord'.[22]

Margaret MacPherson, however, found no support. Although she eventually appended to the Taylor Commission's report a note of dissent in which she forthrightly set out the reasons for her longstanding conviction 'that crofting can never stand on its own feet until the nation owns all crofting lands', there was – realistically – little chance of the Commission following her lead.

Elected to Inverness-shire County Council as Labour member for Portree in 1945, Margaret MacPherson had lost her seat shortly before her appointment to the Taylor Commission. Her defeat was one small indication, among many much larger ones, that the post-war Labour tide was very much on the ebb.

Had Thomas Taylor been reporting to Hector McNeil, he might have been a little more responsive to Margaret MacPherson's point of view. But he was not. By the end of 1951, the Conservative Party had been returned to office. Winston Churchill was back in Downing Street. Neither he nor his Secretary of State for Scotland, James Stuart, as the Aberdeen University Principal well knew, were likely to look kindly on any proposal that croft land should be taken into public ownership.

That was not to say, as the enquiry commission's chairman made clear to his colleagues, that matters should be left entirely as they were. 'The landlord, in disposing of a vacant holding,' Taylor felt, 'should be subject to a certain measure of supervision and control. In our opinion, the situation requires the setting up of an independent authority which, when a holding became vacant or when land in a crofting township became otherwise available, would have the duty of considering how it should be disposed of with a view to maintaining a stable and flourishing crofting community in the township. In particular, it would have to consider whether, and to whom, the holding should be relet as an agricultural unit, or whether it should be attached to one or more of the other holdings with a view to enlargement. In this way, in course of time, the land in many townships could be gradually reallocated and larger holdings created to provide a secure livelihood for those who occupy them.'[23]

Nor should the responsibilities of any 'independent authority' end with the reorganisation of crofting townships, Taylor believed. 'Its main function should be to stimulate the development of the crofting communities in all possible ways.' The new agency, therefore, should have 'powers of land settlement' and 'all grants to crofters, of whatever kind, should be made by it'. It would, of course, compile 'a proper survey and record of croft holdings'. It would also require to be instructed, Taylor thought, to work closely on developmental projects with already established organisations such as the Forestry Commission and the Hydro Board.

'Were such a body created, armed with adequate resources and administered with energy and vision,' Taylor continued, 'we have little doubt that it would obtain an immediate response from the crofting communities, and particularly from their younger and more active members.'

The Crofters Act of 1886, Taylor concluded, had itself resulted in the inauguration of an authority not wholly dissimilar to the body he now had in mind. Its principal task had been to fix fair rents for crofters. It had long ago been disbanded and the bulk of its functions transferred to the Scottish Land Court. But this earlier agency, Taylor stressed, was still remembered fondly in the north of Scotland. It would be no bad thing, therefore, to revive its designation and to provide the Highlands and Islands once again with a Crofters Commission.

The Report of the Commission of Enquiry into Crofting Conditions was published in April 1954. It was not a reassuring document. 'We found many obstacles in the way of full agricultural production from crofting lands,' Taylor and his colleagues commented. 'There were crofts in the hands of absentee tenants, crofts in the hands of old people no longer able to work them fully, crofts which tended to be neglected because their tenants had become absorbed in other employment and crofts which had been taken over as places of retirement by people from other parts of the country. We found amalgamation of crofts sometimes taking forms which militated against full agricultural use. We had evidence of deficiencies in agricultural knowledge, of insufficiency of capital and of the persistence of practices which hampered individual initiative.'[24]

It was 'important', Taylor Commission members stressed, that such problems should be 'fully and frankly discussed'. This they proceeded to do.

Absentee tenants constituted a particular difficulty, the Commission began. 'No adequate statistical information is available to show the extent of this problem, but it exists in varying degrees in all the crofting areas we visited.' Absenteeism had many causes, but it arose most commonly following the death of a crofter whose heir was 'already settled in another occupation elsewhere. Such an heir may have no immediate desire to return to crofting, but at the same time he may feel that the day may come when he will want to do so. He may therefore accept the succession, visit the croft for holidays and eventually return to it on retirement.' Meanwhile, of course, the croft was little used.

Such a situation, the Commission stressed, 'was not envisaged by the authors of the Crofting Acts and . . . it would not have arisen had the original obligation to reside on the holding remained and been enforced. The evidence we obtained in crofting districts suggests that there is a widespread feeling among crofters that the present situation is unsatisfactory. In some districts the position of the absentee tenants was openly criticised. Elsewhere we sensed a reluctance to comment on this issue. In part this seemed to be due to sympathy with the position in which some absentees found themselves. In part it seemed to arise from a reluctance to contemplate any need for change in crofting law.'

By absentees, Commission members were quick to add, they meant only those tenants who lived far away from their crofts. They did not mean those people 'whose ordinary residence is still on their crofts but whose auxiliary employment (for example, in the merchant marine) may take them away from home for months at a time. These latter tenants are part and parcel of the crofting system and the small crofts they normally occupy can usually be looked after by their families. Any agricultural loss which may result from periodic absences in such cases is, we feel, more than counterbalanced by the services which these tenants give.'

Much less easy to defend, in the Commission's opinion, was the fact that many crofts were tenanted by men and women who were simply too old to make any worthwhile use of them.

> The age distribution in many crofting townships has become unbalanced with the result that a very large proportion of the crofts are in the hands of old people. In such townships agricultural production suffers to a marked degree. It becomes difficult to secure the carrying out of the work of the township which must be done in common and a spirit of lethargy tends to prevail. Holdings occupied by old people are, of course, sometimes worked by relatives and sometimes they are sublet for use by other crofters. Such arrangements are not, however, always possible and subletting . . . does not always lead to full use of land. Many holdings remain entirely uncultivated.

Land tenanted by absentees, the elderly and those younger crofters who had clearly lost interest in crofting, the Taylor Commission felt, ought ideally to be redistributed in such a way as to provide larger holdings for people who were still actively involved in agriculture. 'Where the existing crofts are too small and the number of people ready to cultivate the land is declining, it appeared to us that there was everything to be said for some degree of amalgamation, provided it was carried out in an orderly fashion.' But amalgamation of this type was rare, in the Commission's judgment. Even those crofts which fell vacant were seldom reallocated in a genuinely constructive manner. And the blame for this, the Commission believed, belonged to estate managements.

> We had evidence of landlords who had the interest of their crofting tenants very much at heart and who endeavoured to dispose of any holdings falling vacant in the way best calculated to improve the holdings on their estates. On the other hand, we had evidence that on many estates no such policy was pursued. Instances were brought to our notice of crofters, struggling to make a livelihood from small sized holdings, finding adjoining holdings which could have been combined with their own being relet to multiple holders who lived at a distance and who were often of the farming, rather than of the crofting, class. It may be that in such cases the proprietors found it to be to their financial advantage to act in this way. However that may be, the results were clearly disastrous. Instead of an improved holding enabling more land to be cropped, more cattle to be kept and a crofter to make a better living from the land, there resulted too often another holding degenerated into a sheep grazing. Amalgamation in this form involves both the loss of agricultural production and the ruin of a crofting community.

That highly enterprising crofters were to be found in many parts of the Highlands and Islands, Commission members did not doubt. 'We have come across many striking examples of efficient cultivation of crofts and of the skilled management of stock.'

But the obstacles in the way of even the most progressive tenants were immense. The agricultural advisory services provided by the Colleges of Agriculture were frequently inadequate. There was a chronic scarcity of development capital. Agricultural support mechanisms, being national in scope, seldom worked well in the particular circumstances of the north of Scotland. Crofting enterprise was everywhere being impeded by factors beyond the crofting community's control.

The marketing of croft products, for instance, was invariably difficult. These products, Commission members observed, consisted mainly of store sheep and store cattle – young animals which were reared in the townships and on the common grazings but which had to be finished for slaughter on the better land to the south and east of the hills. There were no guaranteed prices for store stock, and it was up to each crofter to dispose of his lambs and calves as best as he could.

> He can send his stores direct to one of the main centres where store sales are conducted, such as Oban or Dingwall, or he can sell them at local sales in the crofting districts to dealers who come out from the main centres. If he sends his stock to a distant market, he has to bear heavy transport charges and he has to accept whatever price may be offered there, since the cost of taking unsold stock home again would be prohibitive. If the crofter relies on local sales, attended by visiting buyers, his problem is simplified. In some areas, however, complaints were made to us that competition among buyers at local sales was at times insufficient to ensure that the crofter obtained full value for his stock. In some island areas there appeared to be difficulty in attracting buyers to come at all – no doubt because of the small number of stock available.

Crofting conditions, the Commission acknowledged, varied greatly from place to place. 'One township will find its chief handicap in the smallness of its holdings, another in the allocation of its arable land, another in the poverty of its soil or the poor quality of its stock. Some are hampered by the mismanagement of their common grazings, others by the absence of shelter, others by the want of drainage or water supply. Some districts are affected most adversely by the state of the township roads, others by the lack of marketing facilities.'

Given these complexities, the Commission concluded, 'no single measure of reform can be regarded as panacea. The remedy, in our opinion, is to be sought in the creation of an . . . administrative organ with wide discretionary powers flexible enough to do what is requisite in the circumstances of each case.'

The 'responsible executive authority' which Taylor Commission members thus advocated in April 1954 was, of course, the Crofters Commission which their chairman had first suggested a year earlier and to which, in the intervening period, a good deal of thought had been given.

The new Commission, Taylor and his colleagues recommended, should be empowered to supervise the reletting of vacant crofts in order to ensure that 'crofts falling vacant in future shall be relet in the way best calculated to promote the interests of the crofting community'.

Because the enquiry team had concluded that 'steps must be taken to abolish the absentee crofter', they were very firmly of the opinion that 'the new authority . . . should be given discretionary power to terminate the tenancy of any crofting tenant who does not ordinarily reside on . . . his holding'.

Those absentees who were thus dispossessed, the Commission of Enquiry suggested, should be allowed to retain the use of their former croft houses and should be permitted to buy from their landlords the pieces of ground on which these houses stood. Similar provision, the Commission suggested, should also be made for those older people whom, so Taylor and his colleagues hoped, the Crofters Commission would succeed in persuading to relinquish their tenancies.

The principal purpose of such measures, the enquiry team stressed, would be to facilitate the 'gradual reallocation' of land from the less active to the more active members of the crofting community.

'We are bound, however, to recognise,' they continued, 'that there are some townships in such a condition of decay that any such gradual process of adjustment would come too late to save them. In these cases, more immediate and drastic action is required . . . We recommend, therefore, that the new administrative authority . . . should have power, on the application of any interested crofters, or at its own hand, to frame a scheme for the reorganisation of any township which has fallen into decay.'

To give the proposed Crofters Commission the authority to re-organise entire settlements and, still more so, to dispossess absentee tenants was, of course, to tamper with security of tenure and, to that extent, to erode basic crofting rights.

But the crofter's position, the Taylor Commission was quick to point out, should be strengthened in other directions. A crofter, for example, should be free to transfer, or assign, his tenancy at any time to any person of his choice – subject only, in certain instances, to the consent of the Crofters Commission. And crofting landholders – instead of being forced to rely on what they could extract from an agricultural support system designed with farming, not crofting, needs in mind – should be provided with a grant scheme formulated specifically to cope with their quite different circumstances.

Nor had the Taylor Commission proved willing to accept the restrictions implicit in a remit which had focussed – more or less inevitably in view of the opinions then current among the civil servants and politicians responsible for framing it – on the purely agricultural dimensions of the crofting problem. It was 'essential that the question of ancillary occupations be gone into thoroughly', Commission members had agreed at their first meeting in 1951. Their published report reflected that determination to take the widest possible view of the crofting community's developmental requirements:

> In the great majority of cases, the croft by itself was never capable of providing a reasonable living for a man and his family. It had to be supplemented by some other forms of work, such as fishing, weaving or knitting. The decline of the crofting system is attributable in great part to the failure of some of these auxiliary industries, notably fishing, coupled with the fact that men and women are no longer content with the modes of life which were acceptable to their ancestors.

Only if fishing, forestry, tourism and other forms of economic activity were expanded in such a way as to provide new job openings for the crofting population, insisted the Taylor Commission, reverting to a theme which had run equally strongly through the Hilleary Report of 1938, would the decline of so many Highlands and Islands communities be halted.

But so precarious was the position in so many crofting localities that remedial action would be anything but easy. 'If the process of decay is to be arrested and reversed, it will require a serious political decision that these communities are not to be allowed to perish; a settled policy well conceived and resolutely maintained for many years; and a substantial expenditure of public money.'

Even if the necessary national commitment were to be forthcoming, the members of the Taylor Commission acknowledged in their report's final paragraphs, it might be asked, as they had asked themselves, if depopulation and dereliction had not 'gone too far to be arrested and reversed. We do not think so. We believe that these communities deserve to be saved from extinction and that they are capable of restoration and survival if the proper measures are taken in time.' It was a brave conclusion.

Within a couple of months of its publication, the Report of the Commission of Enquiry into Crofting Conditions had earned Thomas Taylor a knighthood and – perhaps more important – had elicited from Conservative ministers a promise that action would quickly be taken to implement at least some of the enquiry commission's more central proposals.

'I accept the view of the Commission that the crofting communities ought to be saved from extinction and that conditions of life in those areas can be improved if the proper measures are taken,' the Secretary of State for Scotland, James Stuart, told MPs in April 1954. 'I also recognise that new measures are necessary if we are to secure the full contribution that crofters can make to home food production. I am, therefore, prepared to accept the recommendation that a Crofters Commission . . . should be set up to promote the interests of the crofting communities.'[25]

As was evident from the tenor of these remarks, however, the aristocratic Stuart, for all his very genuine concern for crofters, was as thirled as his Labour predecessors to the notion that only agricultural solutions ought to be applied to crofting problems. 'Successive Governments have rejected the idea of a Highland Development Authority,' ran the notes prepared for the Scottish Secretary prior to Stuart briefing the press on the policy implications of the Taylor Report. What was wanted was simply 'to bring together the various strands of agricultural administration'.[26]

This was not what the Commission of Enquiry had actually said. But it was what the Department of Agriculture, in particular, thought that Sir Thomas Taylor and his colleagues should have said. And it was the Department, of course, which had the primary responsibility for framing the crofting legislation to which the government was now committed.

That legislation, as the Department of Agriculture shrewdly calculated, was not likely to be 'a matter of controversy between the parties'.[27] Labour, having taken no very active interest in crofting during its six years in power, was not strongly placed to criticise the Conservative government's attempts to act upon the recommendations of an enquiry commission which Labour itself had established. Stuart, in the event, was to encounter little difficulty from the parliamentary opposition.

There were, admittedly, one or two slightly querulous voices to be heard on his own side – the Home Affairs Committee of the Cabinet going so far as to suggest to the Secretary of State for Scotland, in the course of the summer of 1954, that it might be 'rather sweeping' to dispossess the absentee tenants of crofts merely on the grounds of their absenteeism.

But Stuart was not to be deflected. 'There is a crying need for the enlargement of crofts,' his ministerial colleagues were informed by the Scottish Secretary, 'and the dispossession of these distant absentees who cannot use their land is one obvious means of restoring the economy of the crofting areas.'[28]

The Crofters (Scotland) Bill – intended, in the words of its preamble, 'to make provision for the reorganisation, development and regulation of crofting' – was accordingly published shortly before Christmas 1954 and received its second reading in the House of Commons towards the end of January 1955.

His Bill, Stuart assured MPs, posed no threat to 'the main principles of the Act of 1886; that is to say, first, the right of the crofter to security of tenure; secondly, his right to obtain compensation for improvements made by himself and his predecessors; and, thirdly, the right to have a fair rent fixed by a judicial tribunal'. These rights would remain in being irrespective of the fate of his proposed reforms, the Scottish Secretary stressed. And far from seeking to 'revolutionise the traditional form of tenure', he sought only 'to strengthen and improve it'.[29]

To this end, Stuart continued, the government were to establish the Crofters Commission advocated by the Taylor Report. The new authority, so the Secretary of State insisted, would be equipped with all the powers it would require to place crofting communities on a wholly new footing. It would control both the reletting of vacant crofts and the transfer of crofts between tenants. It would compile a comprehensive register of croft holdings. It would administer a new set of land improvement grants. It would be able to dispossess absentee crofters and to undertake the reorganisation of entire crofting townships. The net result of all of this, so Stuart confidently forecast, would be a marked change for the better in Highlands and Islands circumstances.

No MP dissented strongly from that prediction. But two struck notably cautionary notes in words which were to prove a good deal more prescient than any used by the Secretary of State for Scotland.

'He personally supported the Bill,' said Jo Grimond, Liberal MP for Orkney and Shetland. But he held out no great hopes for it. And he certainly did not believe that it would 'do very much to repopulate the Highlands or put real heart into the crofting areas'. The Crofters

Commission, Grimond suspected, would have much more limited autonomy than Stuart had implied. The Bill itself made this clear, the Liberal MP claimed. It instructed the Commission 'to keep under review . . . to make recommendations . . . to collaborate . . . to advise the Secretary of State'. This was 'the language of bureaucracy and procrastination', in Grimond's opinion. 'We have been overloaded in the Highlands with representations, advice and recommendations,' said the Orkney and Shetland MP. 'What we need now is some work done.' And he, for one, was not convinced that the Crofters Commission would be up to the job.

Malcolm MacMillan, now 20 years in the House of Commons, was equally pessimistic. 'Crofting is not the whole life of the crofter,' commented the Western Isles MP, taking up one of the themes of the Taylor Report. 'The important point is that the decline of the crofting system is attributable in great part to the failure of the auxiliary industries. Yet we are to give the Commission no powers to go outside the limited field of crofting into the fields where it will have to go if the crofting system as a whole is to be supported, sustained and restored.' Only an agency with the ability to take action on a much wider economic front would succeed, MacMillan felt. Although he wished, he said, to be 'constructive', he saw little prospect of the Commission attaining the objectives set out for it by the government.

On one further point the MPs for the Northern and Western Isles were in complete agreement. It was essential, Grimond stated, that members of the Crofters Commission should be people with a knowledge of crofting. What they did not want, MacMillan added, were 'lawyers, lairds, lackeys or superannuated public servants'.

Neither man, then, could have been greatly reassured by James Stuart's choice of Crofters Commission chairman. The new authority, the Scottish Secretary announced in June 1955, would be headed by Sir Robert Urquhart. This was not a name that meant a lot in crofting circles. Its possessor was a 58-year-old career diplomat who, for the previous four years, had been Her Britannic Majesty's Ambassador to Venezuela.

Should Sir Robert Urquhart have wished, while travelling back to Scotland from South America, to acquaint himself more fully with the nature of his new responsibilities, there was no lack of available reading matter. As Jo Grimond had suggested in the course of the debate on the Crofters Bill, the one commodity of which the Highlands and Islands have always had a more than adequate supply, for the greater part of the twentieth century at any rate, has been advice on how their problems should be solved. Though much of that advice has been rather less than helpful, there was in existence, by the early 1950s, a genuinely informative, and often highly perceptive, literature on crofting.

The Commission of Enquiry into Crofting Conditions had been preceded by Tom Johnston's wartime committees which had analysed various aspects of the north of Scotland economy. The earlier report of the Hilleary Committee still repaid scrutiny. And to those necessarily generalised accounts there had been added the more localised, but extremely well researched productions of the Lewis Association which had been formed in Stornoway in the 1940s in order, as the association itself commented, 'to survey and study the social and economic needs of the Island of Lewis and to draw up progressive plans for development'.

Most interesting of all, however, were the published thoughts on crofting of two immensely intelligent and energetic men who, for different reasons, were not able to contribute nearly as much as they might have done to the betterment of conditions in the Highlands and Islands. One was Adam Collier. The other was Frank Fraser Darling.

Collier, who, like Fraser Darling, had no family connection with either crofting or the north of Scotland, graduated in economics from Glasgow University in 1936 and, shortly afterwards, became principal researcher and statistician to the Hilleary Committee. Having thus been introduced to crofting matters, the young economist continued to think and write about them in the few leisured moments allowed by his war work in industry. But he had still to put his notes into publishable form when, in August 1945, he was killed in a climbing accident in Sutherland.

Some years later, stimulated in part by his own involvement in the work of the Taylor Commission, Alex Cairncross of Glasgow University took in hand the task of editing Collier's material. 'In 1945,' Cairncross wrote of Collier's work, 'it stood head and shoulders above any extant analysis of the social and economic problems of the Highlands and Islands of Scotland.' While it was certainly the case that Collier's posthumously published book, *The Crofting Problem*, had been somewhat overtaken by events in advance of its eventual appearance

The Council of the Scottish Crofters Union at the SCU's 1991 conference in Oban (*Oban Times*)

Angus Macleod, Lewis, the founder of the Scottish Crofters Union
(*Sam Maynard*)

A Lewis crofter outside his newly renovated croft house at Ness
(*Sam Maynard*)

The development of fish farming has created new employment opportunities for
crofters

Lord Sanderson of Bowden, Scottish Office Minister for Agriculture, addresses a
Scottish Crofters Union conference (*Sam Maynard*)

Scottish Crofters Union Director, George Campbell (left), with SCU
Administrator, Fiona Mandeville, and SCU President, Angus MacRae

A wet day at a Skye fank. Alastair Nicolson along with Skeabost crofter Charlie
MacKinnon (*West Highland Free Press*)

Mairi MacIver cutting peat on the family croft at Laxay in Lewis. Mairi's husband,
Iain, was elected Scottish Crofters Union President in 1991
(*The Observer*)

Rogart crofter and postman, John MacDonald, East Sutherland Area President of the Scottish Crofters Union (*Daily Telegraph*)

Feeding sheep in winter (*John Charity*)

Feeding cattle (*John Charity*)

Loading sheep aboard an inter-island ferry at Lochmaddy in North Uist
(*John Charity*)

in 1953, it is still well worth reading. Many of its insights have been cited in these pages.[30]

One difficulty in dealing with crofting, so Adam Collier thought, was that public opinion with regard to crofting issues was commonly founded on 'emotional rather than rational' considerations. 'There are few who could say with confidence what a crofter is, how many there are, or whether all Highland landholders are crofters. The general impression is that crofters struggle individually with their common problems of land scarcity, distant markets and lack of capital in conditions which are virtually uniform and unchanging. Those who can distinguish between the conditions of, say, Orkney and Lewis, or give any account of the economic and social changes of the last half century are in a tiny minority.'[31]

There was 'some excuse for the prevailing ignorance', Collier conceded. 'On many fundamental points it is difficult to speak with any pretension to accurate knowledge. For example, no one has more than a general idea of how many crofters there are or what sources of livelihood they have, or how their standards of living compare with those of other classes elsewhere in Britain . . . . There has been no real assessment of particular resources, far less a survey of total resources. There is not even a reasonably adequate account of the recent economic history of the area.'

These were among the deficiencies which Collier set out to make good. But his tragic death cut short his work. It fell to Frank Fraser Darling to go a long way to completing the task which Collier had so enthusiastically begun.

English by birth and Scots by adoption, Fraser Darling was trained in the Midlands as an agriculturalist. But ever a restless spirit, he quickly abandoned a career in farm management in order to make himself a naturalist and an ecologist. This he achieved not so much by the standard academic means as by spending several years studying deer, seals and other wildlife in several of the more outlying corners of the Highlands and Islands – settling first at Dundonnell in Wester Ross and later living for increasingly lengthy periods in places like the Treshnish Isles and North Rona.

On Tanera Mor in the Summer Isles, where he and his wife Marion made yet another temporary home in the early 1940s, Fraser Darling set out to reclaim and rehabilitate a long-abandoned farm. And it was at this point – influenced, no doubt, by his personal experience of how to set about the business of making a living from the land in so inhospitable an environment – that Frank Fraser Darling began to apply his mind increasingly to the human, as well as to the animal, ecology of the north of Scotland.

'The West Highland problem is one of longstanding,' Fraser Darling wrote on Tanera. But it was a problem, he continued, which had seldom been approached in a spirit of genuinely scientific enquiry. 'Many politicians . . . have given attention to it in moments of political fervour, but not in later periods of administrative office . . . . It is the perfect example of the social problem on which everyone considers himself a pundit but which has received little detailed investigation by anybody.'[32]

That was about to change, however, thanks to the intervention of Tom Johnston who saw in Frank Fraser Darling a man with something of his own capacity to get things done. At the Secretarty of State's instigation, a meeting was arranged and the scientist and naturalist was invited by the Minister to bring the lessons learned on Tanera Mor to the attention of the crofting population.

The result was a series of weekly newspaper articles, published right across the Highlands and Islands, on how to get the best out of a typical croft. And so popular were these notes – on topics ranging from 'The Value of Lime' to 'Care of the Milk Cow and her Calf' – that when they were subsequently gathered together in book form, under the title of *Crofting Agriculture*, the first edition of 5,000 copies sold out within six weeks.

But more than a programme of agricultural education was required to put crofting communities back on their feet, Fraser Darling informed Tom Johnston. What was really wanted was 'a social and biological investigation' of what it was that had caused the crofting problem in the first place. The Scottish Secretary agreed, and in 1944 the necessary finance was found to put Frank Fraser Darling in charge of just such an enquiry. It was to be known as the West Highland Survey, and it was to be conducted – with the help of a team of young, Highland-born researchers – from Fraser Darling's latest home at Kilcamb Lodge, Strontian.

Many hundreds of crofting townships were visited by Frank Fraser Darling and his assistants in the course of the West Highland Survey and the information thus obtained gives a uniquely informative insight into the nature of crofting society at a time when it was being subjected to unprecedented strains. But of still more significance, in retrospect at least, are Fraser Darling's own conclusions as to the ecological aspects of the crofting crisis; for that crisis, he believed, had its origins in the Highlands and Islands having been sapped of much of their productive capacity by prolonged human mismanagement which had begun with the removal of the north of Scotland's tree cover and which had culminated, at the time of the clearances, in the introduction of large-scale sheep farming.

> The Highlands as a geologic and physiographic region are unable to withstand deforestation and maintain productiveness and fertility. Their history has been one of steadily accelerating deforestation until the great mass of the forests was gone, and thereafter of forms of land usage which prevented regeneration of tree growth and reduced the land to the crude values and expressions of its solid geological composition. In short, the Highlands are a devastated countryside and that is the plain, primary reason why there are now few people and why there is a constant economic problem. Devastation has not quite reached its uttermost lengths, but it is quite certain that present trends in land use will lead to it, and the country will then be rather less productive than Baffin Land.[33]

It was possible, Fraser Darling acknowledged, that 'the wilderness value of the West Highlands for the jaded townsman' would be such as to 'justify a large subsidy to maintain a sufficient population of people following practices of misuse to prevent any natural healing of devastation'. But a much more constructive alternative would be to embark on fundamental changes in land use – starting with a massive reduction in sheep numbers.

To bring about the necessary reforms would require 'integrative' action of the sort undertaken by that product of President Roosevelt's 'new deal' policies in the United States – the Tennessee Valley Authority. It was consequently necessary to establish in the Highlands and Islands a TVA-style 'organisation with executive authority' – such an organisation, of course, to have much more extensive economic and planning powers than those which the British Government had bestowed on any existing public agency.[34]

'We wait for the lead to be given,' concluded Fraser Darling. But he was to wait in vain; the Department of Agriculture, committed as it was to a policy of massively expanding the very type of production which Fraser Darling believed to have made a substantial contribution to the degradation of the Highlands and Islands environment, simply choosing to ignore his recommendations.

'Do you know,' Frank Fraser Darling subsequently remarked to a close friend with whom he was discussing his West Highland Survey experiences, 'that I never received an acknowledgment when I sent the final report of the Survey to St Andrew's House? Six years work and not even an official paid postcard in reply!'[35]

Fraser Darling's treatment by the Scottish Office was wholly indicative of the political and administrative establishment's deplorably negative response to a man who was afterwards – in North America, Africa and elsewhere – to make a major contribution to the development of ecological thinking; a man who, perhaps more than any other single individual in the entire post-war period, had both the intellect and the imagination needed to break through the constraints which so surrounded officialdom's handling of Highlands and Islands policy.

But Fraser Darling, as he himself subsequently commented, was 'going dead against' the prevailing 'political . . . trends'.[36] Even when he had become an internationally honoured figure, a Reith Lecturer and a universally acknowledged authority on environmental issues, the Scottish Office was to decline stubbornly to make any real use of his unrivalled experience.

'Successive governments,' remarked Frank Fraser Darling in the 1960s, 'have log-rolled the Highland problem by appointing bodies such as the Highland Panel and the Crofters Commission, given them little or no executive power and placed them under the long arm of the Department of Agriculture which has never had any positive Highland policy.'[37] It was a harsh verdict. But not least in view of what occurred in the wake of the passing of James Stuart's Crofters Act in 1955, it is one from which it is difficult to dissent.

Chapter Four

FALSE START FOR
THE CROFTERS COMMISSION

The Crofters Commission, it had emerged in the early part of 1955, was to have its headquarters in the Highlands – a fact that was thought to be of some symbolic significance in view of the previous tendency for all such agencies, including the Commission's pre-1912 namesake, to be located in Edinburgh. At the new organisation's hastily acquired offices in a somewhat spartan suburban house at 9 Ardross Street, Inverness, Sir Robert Urquhart formally installed himself as Crofters Commission chairman on 1 October 1955.

In addition to its chairman, the Commission had five further members, two of whom had been appointed, like Urquhart, on a full-time basis – the other three being part-timers. To give effect to their decisions, Sir Robert and his colleagues could call upon the services of a small – but soon to grow much larger – team of civil servants headed by the Commission's secretary, Donald J. MacCuish.

There was no lack of work for the Commission to tackle. Nor was there any doubting the enthusiasm with which both commissioners and staff set about their task. But that task, it gradually became apparent, was one of quite exceptional difficulty.

An immediate priority was the compilation of a register of crofts. 'A reliable and comprehensive register,' as the Commission noted, 'is necessary for the efficient discharge of many of our functions and, until its completion, delays must occur in carrying out these functions.'[1] But how exactly was the necessary information to be assembled?

A start was made with the valuation rolls which the various county councils were then obliged to maintain for local taxation purposes. Those properties entered as crofts on the rolls were carefully tabulated. The resulting lists were sent to Highlands and Islands estate managers who were asked to check the Commission's findings against their own rent ledgers.

At this point, perhaps inevitably, all sorts of discrepancies began to emerge. There was dispute, disagreement and doubt as to the status of no fewer than 3,000 holdings. The problems thus caused were made worse by the fact that 'a minority of landlords', as the Commission observed regretfully, 'failed altogether to respond' to the Commission's repeated entreaties for assistance.[2]

It took fully five years to obtain all the details finally shown against the 20,000 or so crofts which were eventually entered on the crofting register. By that point, of course, many of the earlier entries were already out of date.

Still less progress was made by the Commission in the course of its initial dealings with the many 'aged crofters' whose suspicions had clearly been aroused by the Taylor enquiry team's stress on the need to persuade them to make way for younger people. There had been 'widespread misunderstanding' of the relevant legislative provisions, the Commission reported. 'It had been widely rumoured that the Commission were seeking to dispossess aged crofters'.[3]

So persistent was such talk that, in the summer of 1956, Sir Robert Urquhart was obliged to make a public statement as to the Commission's intentions in the matter. 'We have not yet turned any old crofter out of his house,' said the Crofters Commission chairman. 'And we have no case on our hands where such action is likely to arise . . . . The notion that we are going around evicting people is quite, quite exaggerated.'[4]

But if the older residents of Highlands and Islands crofting communities were not to be removed compulsorily, neither were they about to remove themselves voluntarily. In the course of their first 15 months in office, commissioners recorded, they received only five applications from persons interested in the possibility of surrendering their croft tenancies – despite the Commission's constant stress on the fact that elderly crofters could now give up their land without in any way imperilling their rights to continued occupancy of their homes.

Nor did the position improve markedly as the years passed. By the end of 1960 only 70 individuals had expressed any interest in relinquishing their tenancies. And only 40 had actually done so. 'We cannot do more by way of persuasion than we have been doing,' the Crofters Commission stated flatly. 'We think that in this matter . . . both the Taylor Commission and

Parliament overestimated the willingness and capacity of the aged, who constitute the majority of the crofting population, to opt for change.'[5]

The record on absenteeism was a little more impressive. In the course of its first five years, the Commission reported, some 150 absentee crofters had had their tenancies compulsorily terminated. Several hundred more absentees had responded to Commission pressure by giving up their crofts in advance of tenancy termination orders. And many more, seeing the way the wind was blowing, had no doubt taken similar action in advance of any formal prompting from Ardross Street.

Though Sir Robert Urquhart and his fellow commissioners had thus unquestionably demonstrated their determination to root out absenteeism, their standing with the generality of crofters, as the Commission noted somewhat forlornly, had not been enhanced as a result: 'This process is by no means universally popular. We have felt the full measure of the tenacity with which the absentee clings to the family croft, not only the home but the land too, even although he may have little prospect of working it himself. Even in the townships popular sympathy may favour the absentee and oppose the operation of this section of the Act. Township loyalty may count for more than progress, particularly when the majority are aged.'[6]

Much as Crofters Commission members and staff may have wished to enjoy the unstinting support of the crofting population, then, their role, as governed by the Commission's founding legislation, was such as to engender more and more hostility; for, as commissioners all too rapidly discovered, many of their duties were virtually impossible to perform without causing at least some degree of injury or offence.

The Act of 1955, for example, allowed the crofter to assign his tenancy to a person of his choice. But it also made such assignation conditional on Crofters Commission approval. And prior to giving or refusing its consent to a suggested tenancy transfer, the Commission was obliged to take account of the likely impact of the proposed change on the wider crofting interest.

The inevitable outcome was that the Commission, in dealing with disputed assignation cases, of which there were not a few, was bound to aggrieve either the objectors to the prospective tenancy transfer or, alternatively, those parties who had hoped to benefit from it. And the resulting ill-feeling was all the greater in the growing number of cases which involved cash payments of one kind or another.

Though the buying and selling of assignations had not been one of the more widely anticipated consequences of the 1955 Act, it was soon making its appearance. When he hoped to gain financially from such a sale, the Commission observed, 'the outgoing crofter may give scant consideration to the crofting qualifications or suitability of the bidders'. The Commission, however, was required to be more careful – especially if there was known to be local interest in obtaining a croft which, but for Commission intervention, might be made over to an incomer from an entirely non-crofting background. 'We may, in such a case, feel bound to withhold consent,' commissioners reported, 'and thereby incur considerable resentment.'[7]

Such resentment was all the stronger because of the wholly erroneous – but widespread and enduring – conviction that the Crofters Commission could itself direct a particular tenancy to a particular person. In fact, the Commission's powers in this regard were wholly negative. It could reject the nominee of an outgoing crofter – just as it could reject the prospective tenants put forward by the landlords of vacant crofts. In the event of any such rejection, however, it was up to the disappointed landlord or crofter to suggest an alternative tenant. The Commission could not propose its own candidates – a circumstance that was often obscured by the understandable desire of many landowners and crofters to pretend otherwise.

'We have noted with regret,' commissioners complained, 'a continuing tendency on the part of certain landlords and factors to meet the complaints of disappointed aspirants with the entirely misleading statement that the Crofters Commission have "given" a croft to someone else. This is one example of how the Commission are made to serve as a whipping boy. The landlord should, of course, acknowledge that it is he who is responsible for selecting the tenant in the first instance.'[8]

Adding to the Commission's slowly mounting sense of grievance – which is all too evident from the tenor of the passage quoted in the preceding paragraph – was the growing realisation, on the part of its members, that their efforts were failing, more or less completely, to produce the desired regeneration of the crofting areas. Their consequent disenchantment was all the greater because of the inflated hopes which they had harboured in 1955.

The chairman and members of the Crofters Commission had begun by being wildly – and incautiously – optimistic about the new authority's prospects for success. In a manner more than slightly reminiscent of a Soviet commissar unveiling the latest Five Year Plan, Sir Robert Urquhart had personally exhorted crofters to increase their agricultural output by five per cent annually between 1956 and 1959. He was quite convinced, the Commission's chairman told one of the many journalists then beating a path to Ardross Street, that this target would be readily met.

'I am working on the absolutely firm conviction that if the crofter does not get up in the morning it is because he sees nothing to rise for,' said Sir Robert. 'Give him something to aim at and he will rise with the best of them to realise the prospect of a worthwhile life.'[9]

The Commission was 'confident' that, crofters having thus been 'given the chance' so long denied to them, crofting would quickly 're-establish itself as a sane, healthy and independent way of life, one which will repay the effort to preserve it'. There were 'signs' of an 'immense willingness' on the part of the crofting community to 'respond to the promptings of the Crofters Commission'. Indeed it was possible, so Commission members reported, to come to the 'heartening conclusion' that their endeavours had 'evoked unmistakeable evidence of a will to survive'.[10]

'The time is ripe,' remarked Sir Robert Urquhart, 'for restoring prosperity to the Highlands and Islands. Economic conditions are changing in our favour. The fates, so long unkind, have smiled at last upon the crofter; the rain has turned to electricity, the peats to power and bracken can yet become arable. New techniques are at hand when most needed and now the people, too, are promising to come with us.'[11]

In so trying to engender popular enthusiasm for its policies, the Crofters Commission was not without incentives to hold out to crofters. As recommended by the Taylor Report, and as promised by the Scottish Secretary, James Stuart, in the course of the debates surrounding the passage of the Crofters Act of 1955, a Crofting Counties Agricultural Grants Scheme was introduced in July 1956. Its terms, even by the standards of a period characterised by massive public expenditure on farming support, were remarkably generous.

Crofters in more marginal areas – such as the West Highlands, the

Hebrides and Shetland – were entitled to cropping grants of £12 an acre in respect of any land they kept in cultivation. And further grants of up to 85 per cent of the total cost were made generally available for the construction of agricultural buildings and for a variety of farming opera-tions – ranging from fencing and draining to bracken clearance and the reseeding of hill pastures.

The beneficial impact of such measures on crofting agriculture was clear enough. But the overwhelming majority of crofters, as successive governments had been told by successive committees and commissions, did not live by agriculture alone. Although Sir Robert Urquhart, in a radio broadcast following the inauguration of the new grant scheme, expressed the hope that absentee crofters would come home to take advantage of the generous grant aid now on offer, such expectations, like so many others nurtured by the Crofters Commission at this time, proved hopelessly misplaced.

The British people as a whole might never have had it so good, as they were memorably assured in the late 1950s by their Conservative Prime Minister, Harold MacMillan – the grandson, as it happened, of an Arran crofter. But those crofters' grandsons who were themselves still crofters were failing conspicuously to benefit from what was begin-ning to be known as the affluent society.

The building of still more Hydro Board dams, like the provision of the various facilities required in connection with the Atomic Energy Authority's reactor establishment at Dounreay or the army's rocket firing range in Uist, resulted in localised construction booms. But High-lands and Islands unemployment levels nevertheless remained far higher than those prevailing nationally in Britain. The steadily widen-ing gap between north of Scotland living standards and those to be found elsewhere was, beyond doubt, the major cause of a renewed outpouring of people from the crofting areas.

The overall Highlands and Islands depopulation rate was twice as high in the 1950s as it had been in the course of the previous two decades. A number of crofting localities were suffering especially severely. In 1959, as a result, the Crofters Commission, for all its earlier optimism, was obliged to record the existence of no fewer than 220 vacant crofts for which, as the Commission conceded, there was 'no ready demand' and which were consequently expected to 'remain vacant for some time'.[12]

Only the provision of new sources of employment could reverse such trends. That had been among the more notable judgments of both the Hilleary and the Taylor enquiries, and the Crofters Commission quickly reached a similar conclusion. 'All agree that crofting agriculture cannot of itself maintain an adequate and stable population in the Highlands,' the Commission noted. Its members accordingly saw a role for themselves in promoting the required expansion of the regional economy. 'We do think we shall have helped towards a better balance within the crofting counties,' commissioners commented, 'if we can secure for them a measure of modern industrial activity.'[13]

But how was this to be accomplished? There was indubitably an 'apparent lack', as the Crofters Commission observed, 'of any organisation charged with the duty of promoting industry in the Highlands'.[14] But the Commission itself had clearly not been created to make good that particular deficiency – its powers, as government critics like Malcolm MacMillan and Jo Grimond had so cogently pointed out, being confined strictly to the purely agricultural sphere.

In these circumstances, then, all the Crofters Commission could do by way of contributing to the wider development of the Highlands and Islands economy was to appeal to the better nature of those industrialists and businessmen who came to hear Sir Robert Urquhart address a specially convened gathering in Glasgow in the spring of 1956.

'We are a generation of ageing men and women in the crofting counties,' the Crofters Commission chairman told his audience, 'a generation without succession. The young are moving away from us.' For even a sizeable proportion of a community's youth to go elsewhere was one thing, Sir Robert continued. Such movement had always been common in the north of Scotland. 'But now we are at the stage where all are leaving.'[15]

The Commission's basic task was 'to revive agriculture'. And that was being vigorously attempted. The Commission was 'eliminating absentees' and 'negotiating sympathetically with aged crofters' in order to 'enlarge the holdings of active men and women'. But none of that would suffice, in itself, 'to maintain the crofting population'. Industry and jobs were also needed. Sir Robert urged his listeners to provide them.

He had issued those pleas to the Scottish business community, the Crofters Commission chairman noted subsequently, 'to see if they could, from a sense of patriotism, be stirred to branch out into the Highlands'. But his initiative, he admitted, had failed. The men to whom his appeals were directed, as Sir Robert commented on another occasion, might 'express sympathy and understanding'. But in reality, he said bluntly, they did 'not care a hang'.[16]

This was frustrating enough. But what made it doubly hard to bear was the fact that the Commission was also finding it increasingly difficult to implement key components of its agricultural reform programme. Such efforts as had been made to reorganise and redistribute township lands, for example, had served merely to demonstrate that the relevant sections of the Commission's founding Act were a bureaucratic morass capable of bogging down whole teams of hapless administrators for many, many months at a time.

This was precisely what occurred when the Commission became involved, over a three-and-a-half year period, with the crofters of Big Sand in Gairloch. Since nothing is more indicative of the nature of the problems confronting Sir Robert Urquhart and his colleagues than their ultimately abortive attempt to reorganise that single Wester Ross township, it is worth recounting the Big Sand saga in some detail.[17]

At their meeting on 27 August 1956 members of the Crofters Commission decided that Big Sand should be reorganised.

On 5 September 1956 Sir Robert Urquhart visited Big Sand where he met the township clerk and a number of local crofters. Both clerk and crofters expressed their willingness to co-operate with the Commission.

On 7 September 1956 Sir Robert called on the Gairloch estate factor who indicated that, if the Commission were to come forward with reorganisation proposals, these would be considered sympathetically by both himself and his employers.

On 19 November 1956 the Commission's chairman again visited Big Sand. He held a meeting with the township's crofters. He explained the Commission's objectives. And he obtained the meeting's consent to the Commission's proceeding with its plans.

It was now necessary to prepare a list of crofting tenants in Big Sand and so run-down was the township that considerable difficulty was encountered in identifying the tenants of certain holdings. The position was eventually clarified, however, and a list of tenants, following its approval by the Commission on 17 January 1957, was publicised in accordance with Section 19(4) of the Crofters (Scotland) Act, 1955. The accuracy of the published list was questioned by one individual. But his objections to it were shortly afterwards withdrawn.

At their meeting on 23 April Commission members formally confirmed the list of tenants. The confirmed list was duly advertised in accordance with Section 19(5) of the Act. No objections were received.

On 26 April 1957 the Commission's senior lands officer was asked to make a survey of Big Sand with a view to preparing a reorganisation plan. Various visits were made to the township and various discussions held. A draft plan was then submitted to the Commission, the landlord and the Big Sand crofters. This plan was accepted by all concerned.

At their meeting on 12 December 1957 Commission members approved the reorganisation proposals and agreed that these should be incorporated in a draft scheme – not at all the same thing as a draft plan – to be submitted to the Secretary of State for Scotland as required by Section 19(7) of the 1955 Act.

On 30 December 1957, some 16 months after Sir Robert Urquhart's first visit to Big Sand, the Crofters Commission formally approved the draft reorganisation scheme which was then sent to the Scottish Secretary for his approval.

On 17 July 1958 the draft scheme, as laid down in Section 19(8) of the Act, was served on the landlord by the Secretary of State.

In August 1958, and again in February 1959, objections to the draft scheme were received from the landlord. These were the subject of some correspondence among the various parties with an interest in the matter.

On 3 April 1959, some 31 months into these tortuous manoeuvres, the Department of Agriculture advised the Commission that, with certain fairly minor modifications, the Secretary of State for Scotland had approved the Big Sand draft reorganisation scheme.

On 16 April 1959 the Department formally notified the Commission that the draft scheme had been submitted to the Scottish Land Court, in accordance with Section 19(9) of the 1955 Act, for the fixing of croft rents and the assessment of any compensation due to the Big Sand tenants.

On 22 July 1959 the Land Court held the necessary public hearing in the Territorial Army Hall, Gairloch.

On 8 October 1959 the Land Court reported on the new scheme. The report gave details of the rents due on the new – and, of course, larger – crofts which the Commission proposed to create. The report also set out the compensation payments to be made in respect of those permanent improvements, such as fences, which would change hands on the implementation of the scheme.

On 18 November 1959, more than three years after his initial trip, Sir Robert Urquhart again went to Big Sand to discuss the forthcoming reorganisation with the township clerk.

On 9 December 1959 notices containing all the necessary information about the reorganisation scheme were served on the landlord and all the affected tenants, in accordance with Section 19(10) of the 1955 Act. The resident tenants, those notices explained, were now required to vote on the Commission's proposals.

By 9 February 1960 all the Big Sand crofters had registered their votes. Two tenants had voted in favour of reorganisation. The remaining 13 had voted against.

On 11 February 1960, some 42 months after the Commission's first involvement with the place, one of Sir Robert Urquhart's officials called on those Big Sand crofters who, in the end, had rejected the painstakingly planned reorganisation of their township. Seven, he found, had voted negatively because the rents for their new and bigger crofts would be significantly higher than the rents payable on their existing holdings. Two older tenants had decided that they were not, after all, prepared to surrender their crofts. Three disagreed with the way the Commission proposed to reallocate their land. One of the scheme's opponents was not prepared to give any reason for his stance.

Such experiences proved, if they proved nothing else, that the reorganisation procedures contained in the 1955 legislation were, as the Crofters Commission itself concluded, so 'cumbersome and slow' as to be virtually unworkable – all the more so when, as clearly occurred in Big Sand and several other communities, all sorts of powerful social pressures, not least the typical crofter's reluctance to give offence to his elders, operated on the side of those opposing change.

Scheme after scheme foundered in much the same way. By the end of 1960, after more than five years of sustained effort on the part of the Crofters Commission, only five crofting townships had actually been reorganised. In the course of one of its increasingly frequent bouts of introspection the Commission observed:

There is no doubt that too much was expected of reorganisation. It was hailed as a panacea for all the ills of a crofting township in decay. The fact is, of course, that reorganisation can only make a contribution to the restoration of the township by redistributing the land to the best advantage. It cannot by itself provide roads, water, social amenities or ancillary employ- ment and, accordingly, it cannot by itself convert a decayed township of derelict crofts and aged and infirm crofters into a thriving community offer- ing a reasonable standard of living in return for a reasonable effort and with a balanced population of young and old.[18]

Those comments, extracted from one of the Crofters Commission's annual reports, were published during 1959. They seemed to imply that the Commission had effectively abandoned its reorganisation efforts. But nothing could have been further from the truth; for in a memoran- dum which he originally wanted to circulate to Highlands and Islands MPs – but which senior civil servants at the Department of Agriculture advised him to keep to himself – Sir Robert Urquhart was all too evidently considering a more, not less, aggressive strategy. 'We would naturally like to leave our mark on the Highlands,' he wrote, reflecting on the Crofters Commission's first three years in office.[19] Soon both Sir Robert and his colleagues were setting out to explain the drastic means by which they proposed to realise that ambition.

'We saw from our earliest days that many crofters could not make a living without ancillary employment,' commented members of the Crofters Commission in their fourth annual report, published in the summer of 1960. They had consequently sought strenuously to find ways of generating such employment. In its continuing absence, however, they had concluded that the time had come to ensure that crofting agriculture itself made a much greater contribution to the Highlands and Islands economy. That implied reorganisation on a scale never before contemplated. And it meant, in turn, that the Crofters Commission would have to be equipped with much greater powers than those supplied by the Crofters Act of 1955.[20]

A majority of crofters, the Commission declared in the first of a number of passages reflecting its many unhappy experiences in places like Big Sand, were, for one reason or another, no longer working their crofts. 'What is unfortunate is that these inactive crofters . . . rely on their virtual security of tenure to prevent any orderly or comprehensive redistribution of the unworked land amongst those who are capable and willing to work it.' This was a state of affairs, the Commission asserted, which had to be brought to an end.

'As a first step towards the rationalisation of crofting agriculture,' Commission members continued, 'we suggest that ordinary economic trends must be recognised and encouraged.' They had estimated, on the basis of crofting grant returns, that 'about two-thirds of the 20,000 crofts' were not being cultivated. And in conditions of 'normal economic fluidity', much of the more neglected land would have passed permanently 'into the hands of those who need it and can work it'. But the original Crofters Act of 1886 had stopped that happening. 'Security of tenure has frozen crofting agriculture into an outdated pattern of minute units.' It had consequently become essential 'to unfreeze the system'.

The Commission, its chairman and members stressed, had been trying to bring about the necessary reforms. But their powers had proved inadequate.

In a typically decayed township, the majority of the resident crofters have withdrawn from active husbandry and allowed the control of the grazings, and generally of the inbye land as well, to pass into the control of a few active individuals. These latter have a vested interest in the status quo and the following pattern of behaviour has been all too familiar.

At the outset, the crofters concede that their township is decayed and agree that the Commission should draw up a plan for its reorganisation. There ensue operations which may last for up to three years and cost a considerable amount of public money before all the surveys are completed and the final plans submitted to the Secretary of State. These operations inevitably involve much discussion and bargaining with the crofters and the landlord. The crofters may be invited to decide either to renounce the land . . . to maintain the croft as it is or to accept enlargement with the certainty of having to pay a higher rent.

We find that old people, even when they are not working their croft land, dislike taking the decision to relinquish it, while the active men seldom relish the idea of having to pay a rent for land they may already be using rent-free. Even the elimination of absentees tends to set up waves of hostility because the families concerned are known to the resident crofters; and so, as the process drags along its slow, legalistic course, opinion slowly sets against change and the majority inevitably decide to vote down the final scheme.

To help them overcome such resistance, Commission members now suggested, their powers should be greatly enhanced. There would require to be 'further modification of the security of tenure concept'. More stringent controls would have to be exercised over the crofter's right to bequeath or assign his tenancy. The Commission should be given the authority to compel crofters to sublet their holdings if those holdings were not being worked. Instead of being applied to single townships, as in the past, future reorganisation schemes should involve much more extensive districts. And since reorganisation on such a scale would be unpopular, it ought not to be subject to popular votes of the type which had proved fatal at Big Sand.

The overall aim of these reforms, the Crofters Commission made clear, would be to detach from their crofts the two-thirds of crofting tenants who appeared to be agriculturally inactive. That, as the Commission acknowledged, would bring about 'a drastic reduction in the number of croft homes' and, very probably, 'a reduction in population'. But it would also result in the active minority being provided, for the first time, with economically worthwhile holdings.

The crofter of the future, the Commission considered, would require a 'viable unit'. Commission members were in no doubt as to what that implied: 'First, such a unit should not fall short, either as to size or productive capacity, of the type of unit which is regarded as viable in other parts of the country . . . . Second, it should be capable of carrying sufficient stock to yield an attractive income. And third, the tenant should be required to devote his whole energy to working it.'

That a farm should properly be a full-time enterprise, of course, was part of the conventional wisdom of the time. That was why Harold MacMillan's Conservative government, by means of the Agriculture Act of 1957, had gone so far as to introduce grant aid of a type designed explicitly to promote the merging of small holdings. The subsequent Agriculture Act of 1967, passed by Harold Wilson's Labour government, was to make such grants even more widely available.

It was 'desirable that farms which are at present too small to provide an adequate standard of living should be amalgamated into larger units,' declared the Scottish Office in the course of a 1965 assessment of Scotland's economic needs. And statements on the direction of national agricultural policy were no less forthright. One of the 'more important problems' to be tackled was 'that of the small farmer trying to win a livelihood from insufficient land'. The solution, successive governments believed, was to provide such farmers with financial inducements to abandon their holdings which could then be utilised for the enlargement of neighbouring enterprises.[21]

But it was one thing to apply such a policy in more southerly parts of the United Kingdom – where the average holding, even in 1960, was already some 80 acres in extent and where further merging of farms, though it might well result in some rural depopulation, would not produce major social dislocation. It was quite another thing to envisage wholesale amalgamations in crofting districts where full-time farming was exceptional and where many thousands of families still lived on holdings of four, five or six acres.

'If the production of food were to be made the overriding consideration it might point to a general amalgamation of holdings in order to provide more efficient units,' the Commission of Enquiry into Crofting Conditions had concluded in 1954. But there were other, more powerful, considerations to be taken into account in crofting areas, the Commission had continued, and that was why there should be no sweeping attempt to transform crofters into full-time agriculturalists. 'This would be the very negation of security for the present crofting population and would involve social considerations which, we are satisfied, rule it out of account.'[22]

Nor was the Taylor Commission alone in its stance. 'One of the things which has struck me most in my efforts to make a good croft out of a derelict piece of ground is that a croft is not a farm,' Frank Fraser Darling had written in *Crofting Agriculture*. His findings in the course of the West Highland Survey simply served to confirm Fraser Darling in his view that to try to make crofters wholly dependent on agriculture would be to aggravate, not cure, Highlands and Islands ills.

There had been a tendency on some estates 'to unite the land of two or more crofts under one landholder', Fraser Darling noted in the 1940s. 'Such a practice may solve an immediate problem, but it does not maintain the dwindling population.'[23] If such enlargements were to be adopted more widely, Frank Fraser Darling warned, all sorts of difficulties would soon ensue.

Some of the possible social ills consequent on enlargement of holdings . . . may be broadly summarised . . . by stating that in areas of low density of population . . . the density may be so far lowered as to imperil the continuity of the community. Local government becomes loth to spend much on roads; steamers have so little cargo that they cease to call; the numbers of children in the schools get low and the age-groupings ill-assorted; children may be so few that a certificated teacher is no longer provided; the people cease to find fun and games within their own community and there is an accelerated trend to emigration.[24]

The Crofters Commission had initially taken a similar line. In their first annual report, indeed, its members had specifically highlighted the 'dangers and deceptions' inherent in a crofting policy geared to the creation of larger holdings. But there had been 'very much to worry, to puzzle, to disappoint and even to frustrate' the Commission in the five-year period between 1955 and 1960.[25] And now Sir Robert Urquhart and his colleagues had publicly committed themselves to the very strategy they had once denounced.

By so doing they had, in a sense, reverted to a policy first advocated some 75 years earlier. 'To invest the most humble and helpless class of agricultural tenants with immunities and rights which ought to go hand in hand with the expansive improvement of the dwelling and soil,' Lord Napier had observed in 1884, 'would tend to fix them in a condition from which they ought to be resolutely, though gently, withdrawn.'[26] So the Napier Report had recommended that security of tenure be extended only to the small minority of crofters whose livelihood depended more or less completely on agriculture.

That particular recommendation, however, had not been put into effect. The crofting population, through the medium of the Highland Land League, had successfully insisted that each and every crofter, no matter how tiny his holding, should have an absolute right to security of tenure. And what now became obvious, following the Crofters Commission's sudden and unanticipated advocacy of widespread amalgamations, was that crofters – for all that their numbers had been greatly thinned as a result of so much depopulation – were still prepared to rouse themselves to political action in defence of what had been won in 1886.

The Highland Land League had disintegrated in the 1890s, its leadership increasingly embroiled in party political wrangling and its few remaining members wholly disenchanted by the League's failure to

113

build on its earlier achievements. While the magic of the Land League name was such as to make it inevitable that there would be repeated attempts to revive it, the failure of the last of these, in the early 1920s, left crofters with no representative organisation of any kind.

Towards the end of the 1930s, however, several locally-based Crofters Unions were launched to resist projected rating reforms which, had they been implemented, would have had an extremely adverse impact on crofting households. In February 1939, at the height of the controversy generated by the publication of the Hilleary Report, a number of those associations sent delegates to a meeting called by the Lochaber Crofters Union in order to formulate a crofting response to the Hilleary proposals.

The resulting conference – which was held in Mallaig and which was the only such gathering to take place in the Highlands and Islands between the 1890s and the 1960s – was attended by some 30 crofters representing Lewis, Skye, Morar and Arisaig as well as Lochaber. James MacLeod of the Skye Crofters Union was elected conference chairman. Alexander MacCalman of the Lochaber Crofters Union was made conference secretary. The two men afterwards submitted a number of unanimously agreed resolutions to the Secretary of State for Scotland.

Crofters Union representatives, the Scottish Secretary was informed, viewed with 'dismay' the 'steadily increasing depopulation' of the Highlands and Islands; they deplored 'the unnecessarily harsh and difficult conditions under which a large proportion of the population is expected to subsist'; and they affirmed 'their strong belief that action by the Government is immediately required to improve the existing conditions of life in the crofting counties if the drift of population to the towns is to be checked'.[27]

In March 1939 the Secretary of State, John Colville, had talks with an Edinburgh solicitor, Donald Shaw, who was acting on behalf of those Crofters Unions which had met in Mallaig. In the summer of that year Colville had a further meeting with representatives of the Inverness-based Crofters and Smallholders Association – which had been formed a few months earlier and which drew most of its members from Easter Ross, the Black Isle and the northern part of the Great Glen.

But the modest momentum thus generated did not survive the onset of war. No crofting organisation expressed a considered viewpoint on the Attlee government's agricultural policy – for all the substantial impact of that policy on the north of Scotland. Of the various pre-war groupings only the Skye Crofters Union and the Crofters and Small-holders Association, representing between them a very small propor-tion of the total crofting area, were able to galvanise themselves sufficiently to present evidence to the Taylor Commission.

The wider crofting movement initiated at Mallaig in 1939, then, had virtually ceased to exist by the end of the 1940s. To Frank Fraser Darling, for one, this seemed a lapse which would have to be made good if crofting were to prosper.

'It is commonplace to say that the Highlander will not co-operate and that he is rank individualist,' Frank Darling commented. In his experience, crofters, for all that they assisted each other readily enough 'at a sheep-shearing, a potato-planting or a harvesting', found it far from easy to engage in more elaborate forms of co-operation – for either financial or political purposes. 'Simple co-operation and communalistic living is natural to the Gael,' Fraser Darling surmised. 'But the . . . more complex fields of co-operation which reach from the croft to the outside world are more difficult for the township Gael to understand.' And whether or not Fraser Darling was correct in his assessment of the reasons for the crofting community's failure to sustain the means of pressing its own case politically, there can be no disagreement with his conclusion that the survival of crofting, in the second half of the twentieth century, would depend on crofters finding new ways of engaging in joint action on their own behalf.[28]

Farmers had already learned that lesson. The National Farmers Union of Scotland had been formed in 1913 and, to begin with, had expanded only slowly. But NFUS membership had soared during the 1930s and the 1940s – as successive governments intervened more and more in agriculture and as the farming community came to appreciate the value of subjecting politicians and civil servants to intensive lobbying.

Soon NFUS and its sister organisations in other parts of Britain had acquired their enduring reputation as the country's most successful pressure groups. In the decisive post-war period they enjoyed unlimited access to both the Ministry of Agriculture and the Scottish Office. Not only were their senior office-bearers courted assiduously by both the Labour and Conservative Parties; those same office-bearers were given a statutorily guaranteed role in the price-fixing procedures which, from

1947 onwards, were of such critical importance in determining the agricultural industry's overall direction.

The National Farmers Union of Scotland had got off to a bad start with the crofting community as a result of having adopted what the then Board of Agriculture called an 'attitude . . . of direct hostility' to the land settlement measures which were responsible for the creation of so many new small holdings in the Highlands and Islands in the period immediately following the First World War – but which also resulted in a number of tenant farmers, some of whom were NFUS members, being dispossessed to make way for crofters.[29]

And though the rapid growth of farming unions in the 1930s and 1940s resulted in NFUS branches eventually being formed in a number of predominantly crofting districts, such as Lochalsh, Skye, Tiree and Shetland, those branches attracted, for the most part, the comparatively small number of crofters who were full-time agriculturalists. Little effort was made to recruit the mass of part-time crofters. No NFUS branches were established in the Western Isles. The crofting influence on NFUS as a whole remained exceedingly small – so small as to result in the Edinburgh-based union neglecting, on occasion, to make even a token show of defending crofting interests.

In evidence to the Taylor Commission, for example, the Scottish Agricultural Organisation Society recorded its 'concern' that 'no attempt' had been made by NFUS to ensure that more isolated crofting communities, notably those in the Hebrides, could take full advantage of the guaranteed price arrangements which had been introduced some years before. Far too many crofters, the Society believed, were not getting anything like a fair price for their livestock. But there appeared to be little inclination on the part of the National Farmers Union to intervene on the crofting side. 'Unfortunately,' SAOS concluded, 'there has been no crofting organisation, comparable in strength to the NFU, competent to submit and argue the case for the small landholder in northern Scotland.'[30]

But for all the evident need to provide the crofting population with a representative body of the sort implicitly advocated by both SAOS and Frank Fraser Darling, crofters remained almost wholly ununionised throughout the 1950s, their communities in a state of steep decline and their fate increasingly in the hands of a Crofters Commission which, for all its good intentions, was seriously out of touch with crofting

116

opinion and which accordingly ended, in 1960, by advocating a course of action – wholesale amalgamation – clearly considered by most crofters to be contrary to their own best interests.

The crofting challenge to the Commission's amalgamation proposals was launched on 20 August 1960 at a meeting of the Western Isles Constituency Labour Party. The meeting expressed its 'gravest concern at the defeatist attitude of the Crofters Commission,' criticised 'the complete failure of the Commission to press for Government action to bring industry to the Highlands' and condemned 'the emphasis laid by the Commission on the reorganisation of land in dying villages rather than on the urgent steps needed to keep such villages alive'. The meeting also formed a WICLP sub-committee to take the matter further; specifically, to investigate the possibility of forming a Western Isles Crofters Union. To the sub-committee there was appointed, among others, a Lewis primary school headmaster by the name of Charles MacLeod – a fiery and fearless controversialist who, over the months ahead, was to take an all too evident pleasure in castigating the Crofters Commission at every available opportunity.[31]

There was an 'urgent need', MacLeod was soon insisting, 'for the establishment of a Crofters Union to speak authoritatively for crofters'. The entire structure of crofting society, as traditionally organised, was 'in serious danger'. The time had come for crofters 'to organise in their own defence'. Although the Labour Party had taken the lead in ventilating these issues, they required the immediate attention of 'all parties, political and otherwise'.[32]

Rating reform was once more on the parliamentary agenda, Charles MacLeod pointed out. A Crofters Union would obviously serve to defend the crofting community's longstanding derating concessions. 'But we believe that the greatest danger to our way of life, demanding the formation of a union, is contained in the proposals of the Crofters Commission.' These proposals, necessarily involving a huge reduction in the number of crofts, constituted a 'counsel of despair'. Commission members were clearly 'more concerned with crofts than with crofters'. Amalgamation offered no answer to crofting difficulties.

The problems of whole communities whose croftlands are only partly used as a result of the able-bodied having to seek a living away from home cannot be solved by enlargement of holdings. What in reality may be solved are the problems of a very small minority . . . while the problems of the vast majority are aggravated . . . . Let the Commission not be surprised at the anger evoked in the crofting population at the dangers implicit in the Commission's proposals.

Our principle is this, that the population of the Highlands and Islands is worth preserving within the Highlands and Islands and accordingly must be preserved there. Our concern is with people more than land, and it is this principle which the Commission has now betrayed.

We now ask the Secretary of State for Scotland to consider the social implications of the compulsory dispossession of crofters in order to create so-called economic holdings; the ostracism and hostility to which the recipients of such land will be subjected; the ultimate dispersal of the rest of the community to distant places.

The croft is ours by every right upon which a civilised nation bases its social values: by the right of having inherited it as the creation of our forefathers; by the right of having fought for it as our stake in the country at times of war; and by the right of having been born on it and having for it the attachment which everyone has for that spot on God's earth where he was reared. If the Commission can do no better than cast these bonds asunder, they had better realise before they go further that they will earn for themselves the everlasting hatred of crofting folk.

Nothing like this had been heard in the Highlands and Islands since the 1880s. And though Charles MacLeod's remarks were immediately dismissed by the Crofters Commission as 'inflammatory', 'baseless', 'ill-informed' and 'misleading', they clearly struck a chord with crofters. The Commission might – somewhat ludicrously – advise those crofters who were 'troubled or in doubt' to 'let the Commission know' or even to 'turn to someone like (their) minister for guidance'.[33] But the crofting community, for the moment at least, had lost all faith in the Commission and preferred to rally behind the campaign for the formation of an independent union.

A Western Isles Crofters Union was formally inaugurated in Stornoway in January 1961 – with Charles MacLeod, inevitably, as its chairman. A Sutherland Crofters Union had been formed some months before. Other unions were speedily launched in North Uist, Benbecula, South Uist, Barra and Shetland. The Lochaber Crofters Union, which had remained sporadically in existence since the 1930s, was showing renewed signs of life. And so, too, were the Crofters and Smallholders Association and the Skye Crofters Union – the latter of which was able to attract some 400 people to one of its 1961 meetings in Portree.

The Federation of Crofters Unions – with Charles MacLeod, once again, in a key position – was constituted in 1962 in an attempt to give some degree of central co-ordination to the activities of the many disparate local groups. It was one powerful indication of how seriously these developments were taken by already established agencies that the National Farmers Union of Scotland responded to the Federation's emergence by forming its own Crofters Committee and providing that committee's convener with a place on the NFUS ruling council.

The Federation of Crofters Unions was to attract dozens of delegates to its summer conferences, which were held in various parts of the north of Scotland, throughout the early 1960s. Although the membership of its affiliated organisations was never recorded systematically, it is probable that, between 1962 and 1965, several thousand crofters had some link with the revived Crofters Union movement – a link symbolised by the Federation's success in issuing all Crofters Union members with a pocket-sized yearbook and annual report.

Federation deputations were received by Scottish Office ministers. Several Crofters Unions employed solicitors to represent members at Land Court hearings and rating tribunals. Skye Crofters Union candidates were run – successfully – in elections to Inverness-shire County Council. Both individual Crofters Unions and the Federation made authoritative representations on issues ranging from the need for more frequent Hebridean ferry services to the equal need for an improvement in the working conditions of Western Isles tweed weavers.

Levying largely nominal subscriptions and entirely dependent on voluntary effort, both the Federation of Crofters Unions and its constituent associations found it difficult to maintain their initial level of activity. By the later 1960s the Federation – though managing, for the moment, to fend off the moribundity which had so rapidly overtaken its predecessors – was much less of a force than it had been. For all that, however, Charles MacLeod and his colleagues had two major accomplishments to their credit. They had shown that crofters could be unionised and they had made it impossible for the Crofters Commission to proceed with the elimination of the smaller croft.

If the Crofters Commission were to carry out the tasks which Parliament had entrusted to it, the Scottish Secretary, James Stuart, had told the House of Commons in 1955, its chairman and members would require to 'win the trust of crofters themselves'. That was 'vital', Stuart said. The Commission simply had to 'enlist the crofters' own energies'. And if, for any reason, the Commission were not to do so, Stuart forecast, 'success will not crown its efforts'.[34]

That the Crofters Commission, in the event, had failed more or less completely to obtain the crofting community's backing, and that this failure had produced the result which James Stuart had predicted, was all too evident when Parliament again turned its attention to crofting matters – this time to the Bill intended to give effect, in part at least, to those Commission suggestions which had provoked such uproar in the Highlands and Islands.

The new Bill, first debated by the House of Commons in January 1961, the month which also witnessed the formation of the Western Isles Crofters Union, did not give the Crofters Commission the draconian powers which Commission members had requested some months earlier. But the suggested legislation, as drafted, nevertheless went some considerable way to meeting Commission demands – making the case, for example, for accelerated reorganisation procedures and conceding, in addition, that a strengthened Crofters Commission should have the right to compel both absentee and inactive crofting tenants to sublet their holdings to persons of the Commission's choice.

'The trouble was that the procedure for dealing with reorganisation laid down in the 1955 Act was found to be hardly workable,' the Conservative government's latest Scottish Secretary, John Maclay, explained. A tougher approach was consequently needed.[35]

But Maclay had reckoned without the profound suspicion with which the Crofters Commission was now regarded in the north of Scotland – its amalgamation proposals having served only to add to the more longstanding distrust engendered by the way its members had chosen to discharge their administrative duties.

The veteran Labour MP for the Western Isles, Malcolm MacMillan, caught the general mood by stating bluntly that Parliament, far from enhancing the Commission's powers, should be curtailing them. There was, for instance, an urgent requirement for crofters to have 'a right of appeal' to some higher authority on matters such as those tenancy transfers which the Crofters Act of 1955 had left wholly within the Commission's jurisdiction.

He could give 'many examples' of the apparently arbitrary fashion in which the Commission reached its totally unchallengeable decisions on the assignation of croft tenancies, MacMillan told MPs. One of his constituents, for instance, had been informed by the Commission that he was not an acceptable tenant of an island croft because he was 'privately employed on clerical work'. But then the Commission had gone on to sanction the transfer of a similar holding to a civil servant.

'That sort of crude treatment has not helped to improve the relations of the Commission with the crofters with whom it is dealing and with whose problems it is paid to deal,' remarked MacMillan. He went on to make a telling contrast between the Commission's highly secretive proceedings and those of the Scottish Land Court which, the Western Isles MP correctly stressed, was held in the highest possible regard by crofters.

The Land Court, MacMillan declared, 'goes out to the people and it does not insist (as the Commission so often did) on meeting in town halls and civic buildings, but prefers to sit in a local schoolroom, the village hall or anywhere else; to sit down on hard benches and get down to hard facts'. The Crofters Commission, Malcolm MacMillan suggested, might do well to follow the Land Court's example.

This the Commission now agreed to do – issuing a press statement to the effect that its members would henceforth hold public hearings to take evidence on matters which they had previously decided completely in private. But such concessions, predictably enough, came far too late to still the storm of protest which the newly established Crofters Unions were now directing against both the Crofters Commission and the Crofters Bill.

The Conservative MP for Caithness and Sutherland, Sir David Robertson, was in 'no doubt' about the 'resentment' felt by crofters as a result of the way the Crofters Commission had been behaving. In their dealings with crofters, Sir David said, Commission staff 'acted like a prosecution in a criminal trial'. Like Malcolm MacMillan, Sir David continued, he was opposed to amalgamations. But the Commission, in advance of any new legislation, was already promoting amalgamations by underhand and covert means – allowing certain favoured individuals, for example, to accumulate several croft tenancies in localities where less fortunate young men could not obtain a single holding. This was an altogether deplorable state of affairs, and the Commission, Sir David concluded, under cover of parliamentary privilege, were guilty of 'corruption'.

The House of Commons, thundered Malcolm MacMillan, returning to the attack, should not endorse a measure 'which provides for the imposition upon crofters, for periods of several years and at short notice, of arbitrarily selected subtenants'. This would be to dilute and debase security of tenure. It would be to subject crofters to external inter-ference of a kind which would not be tolerated by the Government in the case of farmers or landowners. For crofters to be singled out in such a fashion was quite 'insupportable'.

It was entirely wrong, MacMillan insisted, that legislation should be introduced in order 'to gratify a desire for increased bureaucratic power' on the part of an organisation which was 'responsible for its own failure' and which wanted merely 'to lay new disciplines on crofters'. The Crofters Commission, the Western Isles MP continued, 'is ingrowing and shrinking into itself and is conscious now only of one responsibility and that is to deal with crofting in the narrowest sense. Everything else has been abandoned. The attempt to venture boldly into collaboration with other people in industrial activities has gone.' The Commission, in what amounted to 'a shrill cry of despair' on its part, had been reduced to making the case for compulsory sublets and compulsory amalgamations – courses of action which, if approved, were bound to lead to still more depopulation.

'The Bill,' MacMillan went on, in the course of the longest speech the House of Commons had heard from a backbencher in some 20 years, 'is asking for power, publicly, to discipline and humiliate the crofters for failing to do what no Honourable Member of this House would find himself more able to do than the crofters and that is to wrest a living out of . . . miserable patches of bog.'

It was, observed the political correspondent of The Scotsman, 'a one-man triumph'. Malcolm MacMillan had 'conducted a verbal onslaught of such length and tenacity that he succeeded in making the Govern-ment alter virtually the whole Bill.'[36]

Subletting of crofts would be encouraged. But it would be for crofters themselves to decide if, and to whom, their crofts should be sublet. The Crofters Commission would be required to approve such arrangements.

122

But the Commission would be awarded no compulsory powers – neither with regard to subletting nor amalgamation. All notion of enabling Sir Robert Urquhart and his colleagues to do away with the part-time holding, it was thus made clear, had been irrevocably abandoned.

Chapter Five

THE COMMISSION TRIES AGAIN

The events of 1960 and 1961 forced the Crofters Commission to reappraise both its role and its policies. The latest Crofters Act – for all the mauling it had received from Malcolm MacMillan in the course of its passage through the House of Commons – did eventually result in some streamlining of the reorganisation procedures which the Commission had found so frustrating and which its suggested reforms had been intended to circumvent. But crofting opinion, as articulated by the Federation of Crofters Unions, was now making its influence felt on the Crofters Commission in a quite unprecedented fashion. And crofting opinion was clearly opposed to any further attempts to make full-time farms out of part-time crofts.

Towards the end of 1961, Commission staff wrote to 759 township committees to ask if they might be interested in participating in any future reorganisation scheme. Only 28 replies were received. All of them were negative.

Nor were crofters much more favourably disposed to the officially-sanctioned subletting which was the one significant development made possible by the 1961 Act. 'It is true to say that crofters generally have failed to realise that subletting does not impair their security of tenure,' the Commission noted gloomily.[1] Elderly tenants, in particular, were immensely reluctant to sublet their land. Although the Commission was to continue to bestow its blessing on both sublets and amalgamations designed to produce more substantial holdings, just as it was actively to promote the 'apportionment' or division of common grazings in such a way as to provide crofters with what amounted to enlargements of their crofts, commissioners were now much more inclined than previously to seek community endorsement in advance of promulgating their decisions on such matters.

Greater consultation made for less speedy administration. But it resulted, Commission members reported in 1962, in something of a decline in the overall level of 'contention' surrounding their work. There had been 'a marked decrease', for example, in the volume of complaints about the Crofters Commission received by Highlands and Islands MPs.[2]

The trend towards a more collaborative relationship between the Crofters Commission and the crofting community accelerated in 1963 when Sir Robert Urquhart stood down as Commission chairman and was replaced by James Shaw Grant.

Grant, the owner and editor of the *Stornoway Gazette*, had been a part-time member of the Crofters Commission since its inception. As such he had played his part in the 1960 fiasco and had duly been denounced both by Malcolm MacMillan and Charles MacLeod. But Grant, not least because of his long involvement in the public life of Lewis, an island containing several thousand diminutive crofts, could not but be more appreciative than his predecessor of the key role of the smaller holding in retaining worthwhile numbers of people in localities which, had they been occupied by full-time agriculturalists, would have been almost entirely depopulated. Soon even Charles MacLeod, while continuing to insist that 'the Western Isles Crofters Union was born in bitterness and anger against the Crofters Commission', was acknowledging the extent to which that earlier ill-will had now been 'dissipated'.[3]

MacLeod, forever his own man, had probably debarred himself from consideration for such office by the sheer ferocity of the language he had used in the course of the 1960 conflict. But two other Lewis teachers, John Murdo MacMillan and Alasdair Fraser, both of whom had worked closely with MacLeod in the Crofters Union movement, were appointed members of the Crofters Commission in a gesture intended to set the seal on the rapidly developing rapprochement between the Commission and its erstwhile critics. Since both MacMillan and Fraser had made their names in Lewis as defenders of the part-time holding, there was no mistaking the fact that the Commission was now set firmly on a course very different from that mapped out for it by Urquhart.

The Crofters Commission was still committed to assisting with the overall expansion of the nation's agricultural output, its annual reports made clear. But the Commission, as was stated in the first of those reports to carry James Shaw Grant's signature, no longer considered compulsory reorganisation to be a necessary means to that end. Many crofts would 'inevitably remain small . . . either because of the nature of the land or for social reasons'. There was 'no virtue in amalgamation for its own sake,' and though the formation of 'larger units' might still be undertaken in certain circumstances, such units would be produced only 'by way of voluntary and informal schemes'.[4]

But in seeking to help the part-time crofter, to whose support Grant and his colleagues increasingly pledged themselves, the Commission was greatly handicapped by the nature of the government's approach to crofting. There had been a persistent tendency, Commission members complained, 'to regard crofting as synonymous with agriculture and to restrict schemes for the rehabilitation of crofting to agriculture. It is significant that the generous scheme of cropping and improvement grants for which provision was made in the 1955 Act was not matched by any like provision for any of the crofter's other activities.'[5]

Malcolm MacMillan, among others, had made much the same point some 10 years earlier. Its force was underlined by a 1964 analysis of Crofters Commission expenditure on the Crofting Counties Agricultural Grants Scheme. This showed that the cropping grant – an important element in CCAGS – was going even more overwhelmingly than had been thought to those localities possessing larger crofts and better land. Orkney, for instance, was benefiting much more than Harris; the Caithness parishes of Watten and Wick were faring far better than the Sutherland parish of Assynt.

'There are 245 working units in the Parish of Assynt and 227 in the Parishes of Watten and Wick,' the Crofters Commission reported. 'But the annual average number of claimants for cropping grant in Assynt is only 17, as compared with 137 in Watten and Wick, and the average annual payment of cropping grant is £416 for the Parish of Assynt, as compared with £12,585 for the Parishes of Watten and Wick. Harris, with 535 working units, benefits from cropping grant to the extent of £3,772 per annum, while Orkney, with 532 working units, benefits to the extent of £29,848.'[6]

Such disparities demonstrated the sheer hopelessness of attempting to sustain West Highland and Hebridean crofting families solely by means of agricultural support payments. Other ways of enhancing crofting incomes, the Crofters Commission accordingly asserted, would obviously have to be found. The wider Highlands and Islands economy would have to be expanded, and crofters would have to be provided with much greater opportunities to participate in new forms of economic enterprise.

Many such statements had been made before. But what was striking about the arguments now deployed by James Shaw Grant and his colleagues – most notably Alasdair Fraser – was the extent to which there was implicit, and occasionally explicit, in them a contention that the crofting structure of the Highlands and Islands might, potentially at least, be one of the region's strengths – rather than, as had generally been assumed, its central weakness.

The part-time crofter had always been difficult to categorise. He was not a farmer, but nor was he a fisherman, a labourer, a weaver. He was often all of those things simultaneously. In a massively industrialised country, where a high degree of job specialisation had long been the order of the day, it was consequently tempting to dismiss the crofting tradition of involvement in a wide range of activities as an altogether anachronistic refusal to come to terms with twentieth century realities.

Once the lifestyle of the part-time crofter had been thus condemned, it made sense to maintain, as the Napier Commission had done in the 1880s and as Sir Robert Urquhart had done in 1960, that the crofter should be obliged, in effect, to opt for one status or the other; that he should become either a full-time agriculturalist or, alternatively, abandon his connection with the land and seek employment elsewhere. That had been the pattern in the rest of Britain. There seemed no good reason, to many observers of the crofting scene, why the Highlands and Islands should remain a perpetual exception to the general rule.

Under James Shaw Grant's chairmanship, however, the Crofters Commission began strongly to counter this longstanding view. Bavarian smallholders, the Commission pointed out, were both managing land and working in car factories. Why should crofters not follow a similar route to economic betterment? The crofter, after all, was naturally adaptable. The nature of his life had made him so. On the few occasions when he had been presented with the opportunity of gaining access to it, he had responded well to the factory environment.

'We believe that this willingness of crofters to work in industry, combined with a desire to own their homes and cultivate or stock a piece of land which they regard as their own and which, for practical purposes, is their own, gives us an opportunity in the Highlands of working towards a new form of industrial society which will be healthier and more stable than any community which is completely urbanised,' the Crofters Commission commented in 1966. 'This is the true value of the small croft and, as new opportunities of employment are provided, it will increase rather than diminish.'[7]

Nor was such thinking purely speculative. Caithness and Sutherland crofters, after all, were already working at the Atomic Energy Authority's Dounreay reactor establishment. Lochaber crofters hoped to get jobs at the pulp mill being built on the outskirts of Fort William. And Easter Ross crofters aspired to obtain employment at the massive aluminium smelter planned for Invergordon.

Most parts of the Highlands and Islands, admittedly, were not experiencing developments of the Dounreay, Fort William or Invergordon type. But there were few localities, the Crofters Commission pointed out, which were unaffected by the rapid expansion of the holiday trade. Here, too, was an obvious opportunity for the part-time crofter. 'Tourism will not solve the problems of the crofting areas,' the Commission acknowledged. 'But it is important that crofters should get the largest possible share of this growing industry.'[8]

Because its grant-awarding powers were limited strictly to the purely agricultural sphere, the Crofters Commission could not give financial aid to crofters wishing to embark on new business ventures. That was a matter for the Highlands and Islands Development Board which was established by the incoming Labour Government in 1965 and which, so the Commission hoped, would make a major contribution to the further diversification of the crofting economy.

But the obstacles in the way of such diversification, it was felt increasingly by James Shaw Grant and his colleagues, were not solely of a type to be overcome simply by the HIDB channelling additional public funds into crofting localities. Non-agricultural development, the Commission commented as early as 1963, was 'inhibited by legal and financial difficulties arising from the nature of crofting tenure'.[9] It was in an attempt to find a means of tackling these difficulties that the Crofters Commission was eventually to advocate crofting reforms more fundamental than any which had been contemplated since 1886.

Crofters were the tenants, not the owners, of the land they occupied. Their tenancies, of course, were highly protected – so highly protected that crofters, unlike tenant farmers, were at liberty to establish on their holdings non-agricultural enterprises of the sort that the Crofters Commission was increasingly anxious to encourage.

But because the crofter, for all his freedom to utilise his land as he saw fit, was not that land's proprietor, he remained incapable of raising development capital by the standard means of taking out a commercial loan secured on the value of the asset which his borrowing was intended to finance. Thus a crofter who wished to build holiday chalets could not offer the chalets themselves as security to his banker – because he did not possess an owner's title to the land on which the chalets were to stand and because he could obtain such title, if at all, only by purchasing the requisite piece of territory, at considerable expense, from his landlord.

The crofter who was thus debarred from undertaking non-agricultural development on his own account had no very great incentive to smooth the way for others who might wish to launch new enterprises in his vicinity. Prior to any such enterprise being established on croft land, the land in question had to be purchased by the developer and removed from crofting tenure. While the crofters who had previously occupied such land had to be compensated for its loss, their compensation was assessed only in respect of the land's agricultural value – which was invariably negligible. The very much larger development value was retained by the landlord. It was consequently possible for a crofting township to have part of its common grazings taken over by a commercial concern, for the township's proprietor to pocket many thousands of pounds as a result and for the township's tenants to be entitled to no more than a few shillings apiece.

One such transaction, involving the land required for the construction of the Lochaber pulp mill, was condemned vociferously by the Federation of Crofters Unions in 1966. And at its annual conference the following year, the Federation agreed that the obvious means of righting such wrongs was to permit crofters to become owners of their holdings. 'Crofting legislation is too biased in favour of the landlord in respect of the acquisition of croft land for purposes other than crofting,' the Federation stated. 'The crofter should have the opportunity to buy his croft for reasonable purposes and at a reasonable price.'[10]

In November 1967 this suggestion was put by Federation representatives to one of the Crofters Commission's regular assessors conferences – assessor being the name given to those crofters who, at the Commission's prompting, had been selected locally to act as channels of communication between the Commission and the communities with which it was required to deal. Since the Commission and the Department of Agriculture had already established a confidential working party to examine the 'disabilities under which crofters labour when they seek to undertake non-agricultural development on their crofts or when croft land is required for development by others,' Federation speakers at the conference found the Commission receptive to their proposals – the implications of which, assessors agreed, should be examined urgently by the Commission itself.[11]

The outcome of that examination was made public in the summer of 1968. The Crofters Commission, its members reported, had concluded that many of the provisions of the Crofters Acts of 1955 and 1961 were 'obsolete'. Far-reaching changes were needed, and these changes should ideally result in crofters becoming the outright owners of their crofts.[12]

The case for 'the conversion of crofting tenure into owner-occupation' was 'overwhelming', in the Commission's view. It was bad enough to have 'a specialised and complex legal system regulating fewer than 20,000 holdings'. It was quite unacceptable to allow such a system to impede the orderly development of the Highlands and Islands economy.

'The security of tenure conferred on the crofter in 1886 applies only to the agricultural use of the land,' the Crofters Commission observed. 'He does not share in the increased values arising from changes in land use and he therefore has no incentive to welcome, and co-operate in, development.' Such a situation, the Commission stated bluntly, was no longer to be tolerated.

The Federation of Crofters Unions, the Crofters Commission recalled, had suggested legislation designed to give each crofter the right to choose between tenancy and owner-occupancy. But the Commission had quickly come to the conclusion 'that the only realistic choice was between conversion to owner-occupancy of all crofts . . . or retention of the existing tenure. If there was a piecemeal change, as recommended by the Federation, townships would increasingly become a mixture of tenants and owner-occupiers . . . . This mixture would complicate the administration of agricultural grants and subsidies.'

On an 'appointed day', the Crofters Commission accordingly proposed, all land in crofting tenure should be compulsorily acquired by the Secretary of State for Scotland who would promptly transfer ownership of that land to its occupiers.

Each crofter would thus become the outright possessor of his croft. Each township's common grazing would be vested in a trust which would be elected in much the same way as existing grazing committees. The purchase price payable by crofters to the Secretary of State, who would have already compensated former crofting landlords for their loss, would take the form of annuities which, the Commission hoped, would be approximately equivalent to previous rents and which would run for a fixed term of years.

Although the Crofters Commission, in advocating the abolition of the distinctive form of tenure established by the Crofters Act of 1886, was subsequently to be accused of failing to defend key aspects of the crofting heritage, there was a sense in which the Commission's latest reform package could be presented as an appropriate culmination to processes set in train some 80 years before.

All previous crofting legislation, after all, had been intended to expand the rights of crofters by restricting the rights of landlords. To remove the crofting landlord wholly from the scene, as the Crofters Commission now proposed, was arguably to do no more than reach the logical stopping place on a long, long road. It was also, as the Commission was well aware, to follow the example set in Ireland. And that example, as the Commission also appreciated, was of more than passing interest in the crofting context.

The Irish Land Act of 1881, on which the original Crofters Act was closely modelled by Prime Minister William Gladstone, who had taken a close personal interest in both pieces of legislation, had proved to be no more than a transitional arrangement – Ireland's farmers and smallholders having insisted strongly and persistently that even secure tenancies of the crofting type were insufficient recognition of their claims on the land. British governments had consequently begun, well before the nineteenth century's end, to buy out Irish landlords in order to establish what was then called peasant proprietorship. And though the issue was not finally settled until the 1920s, when much of Ireland had won its independence, both the Irish Free State and the Stormont administration in Northern Ireland were eventually to transfer the ownership of very many thousands of farms and holdings from their landlords to their occupiers by means that were virtually identical to those now advocated by the Crofters Commission.

But if James Shaw Grant and his colleagues, in the course of their efforts to promote crofting owner-occupancy, were happy to refer to the precedents provided by the Free State Land Act of 1923 and the Northern Ireland Land Act of 1925, they were understandably more reticent about previous attempts to persuade crofters to adopt the Irish outlook.

In 1897, following the success of a similar agency in the west of Ireland, the British Government established in the north of Scotland a Congested Districts Board. Its task, as its name implied, was to mitigate the difficulties faced by the crofting occupants of the overcrowded townships which were then characteristic of much of the West Highlands and Islands. The Board was accordingly empowered to ease congestion by redistributing to crofters at least some of the land which, earlier in the nineteenth century, had been cleared of its inhabitants to make way for sheep. Among the properties bought by the Congested Districts Board for this purpose were the Glendale and Kilmuir estates in Skye and the Syre estate in Strathnaver.

Crofters on these estates, including the occupants of the many new holdings created by the Board, were not tenants. They were the owners of their holdings which they were obliged to purchase from the Congested Districts Board by making 50 annual payments – along lines pioneered some years before in Ireland.

But those owner-occupying crofters, unlike their Irish counterparts, were far from happy with their status. As proprietors of land, they were subject to various taxes and charges which tenants escaped. And in 1912, when the Congested Districts Board was abolished, the bulk of its 50-year purchase crofters opted to abandon individual ownership and to become the tenants of the former Board's successor agency, the Board of Agriculture for Scotland – which, like the Congested Districts Board, possessed extensive land purchase and land settlement powers but which, unlike the earlier organisation, was not legally obliged to transfer the ownership of its Highlands and Islands estates to their crofting occupiers.

Only Glendale's 150 or so crofters persisted with owner-occupancy after 1912. But they encountered all sorts of difficulties as a result. Not only were they rated in ways which other crofters were not, they were ineligible – because a crofter was legally defined as a tenant – for the various forms of housing and agricultural assistance which were increasingly available elsewhere in the crofting areas.

The one material advantage of owner-occupancy was that it enabled Glendale people to sell their holdings on the open market. But even that freedom, members of the Taylor Commission concluded following a visit to Glendale in 1952, was likely to lead to 'social and agricultural confusion'.[13] It was certainly the case, by the 1960s, that Glendale did not exactly substantiate the Crofters Commission case that owner-occupancy would result in general advancement – the only Glendale crofters to have benefited markedly from proprietorship being the growing number who had simply sold their crofts and moved away from Skye.

A number of Orkney and Shetland crofters were propelled into owner-occupancy in the 1920s when several Northern Isles landlords sold their estates to their tenantries. But it was more indicative of the overall crofting attitude that, between 1919 and 1945, only nine of the crofting tenants on the many estates in the ownership of the Department of Agriculture – which had succeeded the Board of Agriculture as that organisation had earlier succeeded the Congested Districts Board – chose to exercise their right to buy their holdings.

The ownership of crofts by crofters, Frank Fraser Darling accordingly concluded in 1954, had little to recommend it as a possible 'solution of the crofting problem . . . . Detached inquiry in the Highlands and Islands will show that ownership is not generally desired in the one agricultural sector which enjoys absolute security of tenure; and in the one community [Glendale] where this remedy has been applied, the experiment is breaking down.'[14]

In view of what had gone before, then, it was mildly surprising that the Crofters Commission was able to report, in 1968, that its own local assessors were heavily in favour of a 'total conversion' to owner-occupancy. It was all the more surprising that the Federation of Crofters

Unions was equally enthusiastic – not so much as a single delegate voting against the owner-occupation principle at the Federation's annual general meeting in August 1968.

Were there to be crofting opposition to the Crofters Commission's reform proposals, as historical experience suggested was rather more than likely, the Crofters Union movement, of course, was singularly ill-equipped to act as its vehicle. The Federation of Crofters Unions, after all, had given the owner-occupation idea its first public airing. Federation office-bearers had been closely consulted by James Shaw Grant and other Commission representatives prior to the unveiling of the Commission package. With Federation spokesmen like Charles MacLeod eagerly endorsing Commission policy, and with Commission members like Alasdair Fraser and John Murdo MacMillan continuing to sit on the executive of the Western Isles Crofters Union, it was, in fact, becoming increasingly difficult to make any meaningful distinction, as far as owner-occupancy was concerned, between the position of the Crofters Commission, on the one hand, and the position of the Federation, on the other.

This, in due course, was to lead the Federation into growing difficulty and, indeed, to contribute significantly to its gradual disintegration. But for the moment, there seemed little cause to dissent from the verdict of the Strathpeffer crofter who, replying to a journalist who had asked his opinion of the Crofters Commission's reform package, said simply: 'I think it's one of the finest things that could come to pass.'[15]

James Shaw Grant told another reporter:

In Shetland, I visited a crofter who is busy converting a rather poor part-time unit into a full-time family farm by breaking in hill ground. He has built a beautiful house with his own hands. He has broken in 40 or 50 acres of moorland. He is building a steading. Both he and his wife are working at other jobs to earn the money they are ploughing into the land. But as things stand at present they will never *own* the farm they have created or the house they have built – unless at the landlord's pleasure and at the landlord's price. This is no way to encourage or reward initiative.[16]

Throughout the Highlands and Islands, Grant continued, 'rapid changes' were taking place 'in the demand for land, the uses to which it is put and the price which it can command . . . If one were deliberately seeking the most inefficient method of ensuring good land use in such a situation it would be difficult to invent anything more cumbersome than the elaborate system of checks and balances which has grown up since 1886.' It was this system which the Commission was now determined to simplify. 'The four essentials are to enable changes in land use to take place more freely; to give the crofter an incentive to accept them when they are in the public interest; to give him the opportunity of initiating them when they are within the capacity of a private individual; and to reduce controls to a minimum.'[17]

Crofting proprietors, needless to say, were openly critical of the Commission's strategy. 'These proposals have been drawn up from the crofting viewpoint only,' said a spokesman for the Scottish Landowners Federation. And the SLF's Highland members were still more forthright. He objected strongly to 'the threatened fragmentation' of crofting estates, remarked Lord Burton of Dochfour, 'as not in the interest of efficient farming and possibly disastrous for landowners'.[18]

The press, however, was more sympathetic. The Crofters Commission, commented *The Times*, had devised a workmanlike solution to the 'vexed question' of how to modernise crofting. *The Scotsman* agreed:

> For more than a decade, the Crofters Commission has been wrestling with an intractable problem. Its work has not been entirely barren; it has channelled a good deal of aid to crofters and encouraged some land improvement schemes. But the Commission has failed in its main purpose of making crofting a viable and prosperous way of life. Many crofts are unused or underused and the Commission's efforts to reorganise townships have usually foundered on local resistance. It is not the Commission's fault that its labours have been unrewarded. It has been baffled by the complex legal system relating to crofts.

But the Commission, *The Scotsman* concluded, had at last 'produced proposals for radical reform'. And the case for such reform, in the paper's view, was 'overwhelming'.[19] That was, indeed, the general view – with only two publications striking any kind of cautionary note with regard to the Commission's plans. *The Economist*, while endorsing the Commission's overall approach, warned darkly that 'wild Highland rows' would 'doubtless now arise'. And the *Inverness Courier* was even more pessimistic. 'If there is one thing certain about the Crofters

Commission proposal that there should be new legislation enabling crofters to become owners,' the newspaper predicted, 'it is that its first fruit will be controversy.'[20] And so it proved.

At a series of public meetings organised by the Commission during the winter of 1968–69, owner-occupation was generally supported. But for all the Commission's efforts to ensure unanimity, more and more discordant voices were beginning to be heard.

A Skye crofter thought that the Crofters Commission would end by splitting the crofting community in two. A Sutherland crofter felt that owner-occupancy would bring 'an awful lot of problems'. An Inverness-shire crofter warned that entire communities were about to 'be stripped of the protection of crofting tenure and its many attendant benefits'. Of the 114 crofters voting in a referendum organised on the Dunvegan Estate in Skye, only 32 favoured owner-occupation. And at a noisy gathering in Stornoway, the Crofters Commission's chairman was subjected to a 'vigorous attack in Gaelic' by an Aignish crofter who 'questioned Mr Grant's fitness for the post and accused him of trying to do away with crofting'.[21]

A North Uist minister, Revd James Morrison, told his congregation that crofting tenure had served them well and should not be lightly given up. The then Moderator of the General Assembly of the Church of Scotland, Revd Dr Thomas Murchison, whose close connection with the politics of crofting had begun with his involvement in the Highland Development League of 1936, was even more forthright. 'I can see this change in crofting tenure leading to a second wave of clearances,' Murchison remarked. 'It is very nice for present crofters who become owners of the land overnight and tomorrow can sell out at any price,' the Moderator added. 'But young people from crofting localities would not be well served by a land market in which crofts went simply to the highest bidders.'[22]

Tom Murchison was not alone in his stance, acknowledged Charles MacLeod of the Federation of Crofters Unions. A common criticism of the owner-occupation proposal, wrote MacLeod in August 1969, 'is that crofters would sell out to the demand for holiday homes and that therefore a unique way of life would be gone'. But such criticism was unacceptably patronising, MacLeod insisted. Underlying it, in his view, was 'the belief that crofters as a class cannot be trusted to manage their own affairs without control by landlords, factors and officials'.[23]

But Charles MacLeod's former ally, Malcolm MacMillan, was not convinced by such reasoning. The Crofters Commission had produced plenty of arguments in favour of a move to owner-occupation, MacMillan commented in the course of what was to be one of his last speeches in the House of Commons. 'But I am extremely worried,' said the Western Isles MP, 'about only one side of the case being effectively and continuously put, since there is no body parallel to the Crofters Commission, with all its resources and authority, to put the other side of the case.'

The Commission made much of the need to enable crofters to raise commercial capital, MacMillan continued. But the owner-occupier of an island croft, in his opinion, would not be much better placed in this regard than a crofting tenant. 'The bank will not look at his little four acres and say, "This is an excellent thing to be offered for security." ' The Crofters Commission, with the apparent backing of the Federation of Crofters Unions, MacMillan concluded, was proposing to 'steam-roller out of existence all the legislation right back to 1886'. That would be 'a most serious undertaking'. And it was 'something which should not be done in a hurry'.[24]

As a Labour MP representing the major crofting constituency, MacMillan was bound to have some influence on Harold Wilson's Labour Government to which James Shaw Grant and his colleagues were looking for the legislative measures necessary to implement their proposals. And though the Commission thought it 'essential, in the interests of crofters and crofting, that the period of uncertainty in regard to the Government's intentions should be as short as possible,' it was clear that the Labour administration's Scottish Secretary, Willie Ross, was not hastening to make up his mind. There was 'no simple or ready solution to the crofting problem', Ross remarked in November 1969.[25] Further consultation was required. And that consultation was still in progress in the summer of 1970 when Harold Wilson rashly called the general election which resulted in Edward Heath's Conservative Party gaining power.

The force of the Crofters Commission contention that existing tenurial arrangements were outmoded was dramatically underlined in Skye in 1972 when an island landlord, Horace Martin of the Strollamus Estate, applied to the Scottish Land Court for permission to remove from crofting tenure a two-acre slice of common pasture on which he intended, according to his lawyers, to erect eight houses. Martin's application, though opposed strongly by his crofting tenants, who drew national attention to their case by staging an anti-landlord demonstration of a type not seen in Skye since the 1880s, was granted by the Court. And though the resulting house sites were reckoned to be worth more than £1,000 apiece, the compensation due to the eight affected crofters amounted to only £1.50 each.

This was, of course, exactly what the Commission had sought to prevent by suggesting that crofters should acquire those ownership rights which resulted in landlords being the sole financial beneficiaries of land use changes of the Strollamus type. Conservative Ministers went some way to accepting the Crofters Commission case in August 1972 when the Secretary of State for Scotland, Gordon Campbell, announced the Government's intention to give crofters the right to buy their holdings.

But for all that Campbell labelled his proposals a 'charter for crofting freedom', there was no disguising the fact that they fell a long way short of the Crofters Commission's 1968 demands.[26]

There was to be no general and compulsory transfer of ownership. All crofters, admittedly, would be entitled, in future, to share in development values of the sort which Horace Martin and other landlords had previously monopolised. All crofters would have a right of purchase which, failing agreement between tenant and landlord, would be enforceable by the Scottish Land Court. But purchase prices would themselves reflect the development value of the land in question and would, therefore, be considerably higher than the Crofters Commission had originally forecast. And because each and every crofter would be entitled to exercise, or refrain from exercising, his right to buy, townships were bound to be fragmented in ways which the Commission had earlier warned against.

That Gordon Campbell's vision of the crofting future should have been described as a 'considerable improvement' by the Scottish Landowners Federation was readily predictable. What was more surprising was the reaction of the Crofters Commission. Although the Government's intentions were 'different' from their own, Commission members announced, the proposed reforms would, in their opinion, 'represent a really substantial improvement in the crofter's position –

the most important, in fact, since the first Crofters Act of 1886'. Accordingly, the Commission would commend the Campbell plan to crofters.[27]

The Crofters Commission, however, was no longer having things all its own way on the owner-occupancy issue. A number of the Labour Party's more prominent activists in the Highlands and Islands – notably Margaret MacPherson, the Skye crofter who had served on the Taylor Commission, and Allan Campbell MacLean, an Inverness author who had been Labour's unsuccessful candidate for Inverness-shire in the general election of 1966 – had been openly hostile to the owner-occupation concept from the outset. As early as the spring of 1969, they had succeeded in getting Labour's Scottish conference to adopt a motion which both condemned Crofters Commission policy and advocated the transfer of Commission functions to the Highlands and Islands Development Board. By 1972, when Gordon Campbell launched his crofting policy, MacLean and MacPherson had been provided with a ready platform for their views by the newly established – and intensely radical – Skye newspaper, the *West Highland Free Press*.

The Scottish Secretary's 'charter for crofting freedom', Margaret MacPherson informed *Free Press* readers, gave 'no freedom . . . to the crofter unless it be the freedom to increase the rate of decay in crofting townships.' It was 'astonishing', Allan Campbell MacLean added, that the Crofters Commission should have given its blessing to a reform package which, in several key respects, 'flatly contradicted' the principles which Commission members had themselves laid down some four years earlier.[28]

Labour criticisms of owner-occupation, Charles MacLeod had steadfastly insisted on behalf of the Federation of Crofters Unions, were no more than the babblings of 'dilettante socialists'. Crofters themselves, he added, were 'almost unanimous' in their backing for reform.[29]

But for all MacLeod's efforts, there were growing signs that the Federation no longer spoke for crofters on the owner-occupancy question. At an unexpectedly stormy meeting of the Skye Crofters Union in November 1972, for instance, the Crofters Commission was roundly condemned for 'having blundered into' a situation likely to result in crofters being subjected to 'landowners' legislation'. Soon the leaders of the Skye Crofters Union and the Western Isles Crofters Union were publicly at odds on the merits of the Conservative government's crofting proposals – with Charles MacLeod and his colleagues in Lewis continuing to endorse them in the face of Skye comments to the effect that they should be rejected.[30]

'There is little leadership from within the crofting community and no statesmanship whatsoever,' lamented one Lochalsh crofter. The further furore surrounding the Labour Party's formal response to the Campbell plan – a response which took the shape of a policy paper calling for Highlands and Islands estates in excess of 4,000 acres to be taken into the ownership of the Department of Agriculture – would not have caused that man to alter his assessment. 'Perpetual serfdom under the Department is not many crofters' vision of the millennium,' Charles MacLeod said bitterly.[31]

The Secretary of State for Scotland, meanwhile, was pressing ahead with legislation and, by the start of 1974, was able to present a Crofting Reform Bill to the House of Commons. He hoped, Gordon Campbell told MPs, 'to offer crofters fresh opportunities'.[32] But before the Scottish Secretary's intentions in this regard could be put properly to the test, Campbell, the Crofting Bill and, indeed, the Conservative government had all fallen victim to the electoral consequences of Edward Heath's decision to go to the country in the hope of obtaining a mandate to confront the National Union of Mineworkers. Labour was once more in office, and crofting reform was back in the melting pot.

But neither Harold Wilson's return to Downing Street nor the reinstatement of Willie Ross at the Scottish Office were to result in the radical new departures for which Labour activists in the north of Scotland had been pressing.

The Scottish Council of the Labour Party had set up a working party to consider crofting policy. Its members included Allan Campbell MacLean, Margaret MacPherson and Brian Wilson – the latter being the editor and principal proprietor of the *West Highland Free Press*. Its conclusions, unsurprisingly, were anything but favourable to owner-occupation – MacLean, MacPherson, Wilson and their colleagues coming out strongly in support of a form of community ownership which, the working party maintained, could be brought into existence by making all croft land over to an elected Crofting Trust.

The possibility of vesting the ownership of crofting properties in community trusts, ironically enough, would eventually be explored by Margaret Thatcher's Conservative Ministers in the later 1980s. But it did not appeal to Willie Ross who – whatever he may have thought of Charles MacLeod's description of Allan Campbell MacLean and Brian Wilson as 'doctrinaire socialists' attempting to thwart 'the final emancipation of the crofting people' – was clearly more inclined to take advice from the Crofters Commission than he was to pay attention to the views of the Labour rank and file. The government would proceed to legislate on crofting, Ross duly announced. And despite the fact that the owner-occupation principle had been rejected by more than one of Labour's Scottish conferences, the Labour Secretary of State, it appeared, intended simply to revive, and slightly amend, the Bill devised by his Conservative predecessor.[33]

'The Government,' said the Secretary of State, making clear his own position, 'have no plans at the present time for bringing all land within public ownership and the acquisition of crofting estates is not contemplated as a separate measure.' Neither community ownership nor land nationalisation, in other words, were on Labour's political agenda.

But he was 'keenly aware', the Scottish Secretary commented, on introducing his own Crofting Reform Bill to the Scottish Grand Committee, that 'many crofters' were anxious to buy their crofts. And Labour's legislation was intended to facilitate such purchases.[34]

A crofter, Willie Ross continued, would henceforth have an incontestable right to buy both his house site and the surrounding 'garden ground'. He would have the further right to acquire the ownership of all, or any part, of his croft land. And, Ross stressed, the price payable by a crofter wishing to acquire ownership of these assets would be determined solely by the agricultural value of the land in question. The land's development value, which the Conservative government had proposed to build into the purchase calculation, would not be so included by Labour. 'If the land has any value,' Willie Ross remarked, 'the person who has created that value has been the crofter.'

In recognition of that principle, the Secretary of State went on, he proposed, as the Conservatives had also proposed, to give even tenant crofters a share in any development value which might be realised on the sale of croft land, including common grazings, for non-agricultural purposes.

Thus there was eventually tackled the particular grievance which, some 10 years earlier, had first led the Federation of Crofters Unions to ventilate the possibility of crofters being permitted to buy their croft land. But if the united crofting front created in the aftermath of the events of 1960 had fragmented in the course of the owner-occupation argument, Scotland's politicians, or the more senior of them at any rate, were happily at one on crofting policy. 'This Bill is substantially the same Bill as we debated in January 1974,' Willie Ross was informed by Hamish Gray MP, the Conservative spokesman on Highland affairs. 'Therefore, I . . . congratulate the Right Honourable Gentleman for having had the courage to go ahead with the Bill in face of the opposition to it shown by sections of the Labour Party in the Highlands of Scotland.'

The Crofting Reform Act of 1976, said Willie Ross, was 'a watershed in crofting history'. The Crofters Commission, which had campaigned so hard to obtain the new legislation, unsurprisingly agreed. The passing of the Act, the Commission observed, 'marks the beginning of what may possibly be the most important period in the evolution of crofting since 1886'.[35] In the event, however, all such expectations – as experience might have suggested to those who expressed them – were to prove exaggerated.

Crofters benefited considerably, of course, from those legislative provisions which entitled them, in effect, to half of the development value of any tenanted croft land which was taken out of crofting and put to other uses.

When, in 1977, a five-acre plot in Uig, Lewis, was removed from crofting tenure and sold by the local landlord to a developer, no less than £10,000 was paid to the crofters concerned – a sum which contrasted very markedly with the £3 which crofters in the same district had received in the wake of a similar transaction in 1966. With croft land in steadily growing demand as a result of developments such as the rapid expansion of fish farming, there were to be many more such payments made to crofters in the course of the 1980s.

Nor were such payments the only positive outcome of the reform legislation. The ease with which a crofter could now acquire full legal title to his house site, and thus gain readier access to mortgage finance from banks and building societies, was generally welcomed – as was the fact that the 1976 Act gave common grazings committees new powers to initiate township improvement schemes.

Less warmly received, however, were those aspects of the Reform Act which, it gradually became apparent, were to have the effect of reducing the number of crofts and limiting crofting access to grant aid.

A holding purchased by its occupant did not cease, by virtue of that fact alone, to be a croft. Its previous tenant, to be sure, was now its landlord. But that did not make him any less subject to the dictates of crofting legislation.

Nor did the mass of crofting law which had been enacted since 1886, and most of which was unaffected by the 1976 reform measure, make any distinction, in principle, between the landlord of a single croft (in other words, an owner-occupying crofter of the sort the new Act was intended to bring into existence) and the landlord of many crofts (in other words, an estate proprietor of the traditional type). Because his holding had become untenanted, as a result of his buying it, the owner-occupier was, in strictly legal terminology, the landlord of a vacant croft. And because earlier legislators had expended considerable effort on devising means of ensuring that the landlords of Highlands and Islands estates could not gradually remove those estates from crofting tenure by allowing vacant crofts to remain vacant, the owner-occupier, in accordance with the statutory provisions dealing with the landlords of vacant crofts, was, in principle, liable to be instructed by the Crofters Commission to instal a tenant on the holding he had just purchased.

In practice, of course, the Commission did not issue reletting orders of this type to owner-occupiers who remained resident on their crofts. But should an owner-occupier wish to realise the value of his newly-acquired asset – by offering his holding, for example, to a company in search of land for commercial afforestation of the sort which became more and more common in the north of Scotland in the years immediately following 1976 – then he was very likely to be advised by his lawyers to remove his croft from crofting tenure prior to putting it up for sale. No forestry company, after all, was going to spend good money in order to become the landlord of a piece of ground which the company, unlike the owner-occupying crofter, might well find itself forced to let to a new crofting tenant.

The possibility that owner-occupation would lead, in just this fashion, to a marked contraction in the total area of croft land had been one of the points made repeatedly by Brian Wilson, Allan Campbell MacLean and others during the early 1970s. The 1976 legislation had accordingly been framed in such a way as to oblige the Crofters Commission, prior to its granting the decrofting order which alone could remove a holding from crofting tenure, 'to have regard', in the words of the Commission's own *Guide to the Crofting Acts*, 'to the general interest of the crofting community in the district in which the croft is situated and, in particular, to the demand, if any, for a tenancy of the croft from a person who might reasonably be expected to obtain that tenancy if the croft were offered for letting'.

Those provisions could arguably have been interpreted by the Crofters Commission in such a way as to make decrofting very difficult. But the Commission, having been the principal author of reforms which its chairman and members had canvassed strongly on the basis that they would bring financial gains to crofters, could not very plausibly do anything other than facilitate a process calculated to result in owner-occupying crofters profiting handsomely from their new status. Soon crofts were being decrofted at an average rate of one a week. The consequent controversy was all the greater because many larger holdings were involved and because the Crofting Reform Act – in a clause which, though it attracted little attention in the course of the legislation's passage through Parliament, was afterwards to be widely condemned – made it legally impossible to create new crofts to take the place of those which were being lost.

Crofters who bought their holdings with a view to decrofting and selling might have been regarded with some disfavour by those members of the wider crofting community who were angered by the threat which such practices clearly posed to the long-term continuation of crofting. But for all that he might be criticised by other crofters, the owner-occupier who adopted such a course had obviously benefited from the acquisition of his holding. That was more than could be said about many crofters who, having become owner-occupiers, chose simply to get on with working their land.

As early as 1970, Alasdair MacKenzie, who was then the Liberal MP for Ross and Cromarty and who had previously been a member of both the Taylor Commission and the Crofters Commission, warned that owner-occupation might result in reduced entitlement to crofting grants of the type introduced by the Crofters Act of 1955. 'Those benefits are given to crofters,' MacKenzie remarked. 'And once you abolish the crofting system and put all small units throughout Scotland on the same footing there is no guarantee that Governments will continue the present benefits.'[36]

The Ross and Cromarty MP was promptly told by the Crofters Commission that he had 'fallen into . . . error'. And the Commission's chairman, James Shaw Grant, was subsequently to comment categorically: 'Grants and loans for crofters will not be affected by the proposals for tenure reform.'[37] But on this point James Shaw Grant was wrong.

As Alasdair MacKenzie stressed more than once in the course of the owner-occupation wrangle, he himself had been largely instrumental, during his time with the Crofters Commission, in drawing attention to the injustices arising from the fact that the Crofting Counties Agricultural Grants Scheme, as originally implemented in 1956, was legally confined to the tenants of registered crofts. The assistance to be got from the CCAGS was consequently unavailable to owner-occupiers of the Glendale or Orkney type. The Crofters Act of 1961, following Alasdair MacKenzie's prompting, went some way to right this wrong by extending CCAGS benefits to any Highlands and Islands smallholder who could demonstrate, to the satisfaction of the Department of Agriculture, that he was of 'like economic status' to the generality of crofters.

The 1961 reform, as intended by MacKenzie, thus had the effect of extending eligibility for CCAGS aid. The 1976 Act, in contrast, was sharply to reduce such eligibility – as Alasdair MacKenzie had predicted.

What the Ross and Cromarty MP grasped much earlier than the Crofters Commission – which seems not to have become properly aware of the force of his argument until the 1976 legislation had actually reached the statute book – was that anyone who chose to purchase his croft would then have to submit himself, in effect, to a means test in order to obtain financial benefits which he had previously got without question.

Nor were these proceedings a mere formality. Crofters who had bought their crofts in the hope of liberating themselves finally from the lingering remnants of the controls once exercised so harshly by crofting landlords now found themselves having to supply Department of Agriculture civil servants with details of every penny earned by themselves and their wives. Since a married couple's total income did not have to be very high to debar them from crofting grants, owner-occupiers, to their understandable anger and resentment, found themselves cut off completely from assistance which was still available automatically to other – possibly much more affluent – crofters who chose not to exercise their purchase rights.

After 1976 any crofter was free to acquire his croft from his landlord for a sum equivalent to 15 times his holding's fair rent. Since such rents were generally low, most crofts could be bought for under £300. There was, however, no rush to take advantage of these terms.

Decrofting disputes; the operation of the 'like economic status' regulations; the fact that an owner-occupying crofter, even if he still qualified for CCAGS assistance, quickly ceased to be eligible for crofter housing grant and loans: all had the effect of rapidly discrediting the purchase principle. A petition expressing 'determined opposition' to the Crofting Reform Act was circulating among Lewis crofters within a few weeks of the measure having passed into law.[38] And though there was no chance of the petition's organisers attaining their objective, the Act's immediate repeal, their openly expressed hostility to the owner-occupancy concept was soon to become the dominant crofting attitude.

'As a piece of legislation,' the Crofters Commission itself acknowledged in 1981, 'the 1976 Act was grafted on to, rather than integrated with, the older crofting legislation and some minor flaws have become apparent.' Three years later the Commission had become even more disillusioned with the legislation it had done so much to promote. The purchase of crofts, admitted the Inverness solicitor Jeff MacLeod, who had succeeded James Shaw Grant as Commission chairman in 1978, had 'ground to a halt'. Crofters, MacLeod added, were 'not prepared to be means-tested on account of becoming owner-occupiers'. Among those who had opted for owner-occupancy, the Commission's chairman felt obliged to agree with a quizzical journalist, were 'some' who had 'lived to regret it'. Crofters who were still tenants, MacLeod said, would be 'ill-advised' to buy their holdings.[39]

Even by the end of the 1980s, nearly 15 years after the enactment of the right to buy, only some 15 per cent of all the 17,500 or so crofts on the Crofters Commission's crofting register were shown as owner-occupied. Most of these were in Easter Ross, Caithness and the Northern Isles. Of the 5,981 crofts in the Western Isles, the principal crofting heartland, only 44 had been purchased by their occupants.

147

This was not much to show for a reform in which the Crofters Commission had invested so much of its credibility. And it was, perhaps, of some significance in this context that, after 1976, the Commission was never again to embark on far-reaching initiatives of the type which had led it, first, into the amalgamation debacle of 1960 and, second, into the scarcely less debilitating quarrels surrounding the 1976 legislation.

As Crofters Commission chairman, James Shaw Grant, like Sir Robert Urquhart before him, had been both full-time and highly visible. Jeff MacLeod, however, was a part-time chairman who was apparently content to confine himself largely to administrative matters and who steadfastly refrained from making sweeping statements on the crofting future. If an organisation like the Crofters Commission could be said to have had a heroic age, then that heroic age was now well and truly over.

Chapter Six

THE HIGHLAND BOARD
TACKLES THE LAND QUESTION

The Crofters Commission's owner-occupation proposals had been framed, in part at least, in response to the establishment in 1965 of the Highlands and Islands Development Board. But the new Board itself, or so it was widely believed in the north of Scotland, was likely to want to make its own distinctive mark on land management issues. This was not an expectation which the HIDB's founders did anything to discourage.

A development agency had been long awaited in the Highlands and Islands. Seriously considered by Scottish Office ministers in the 1930s, it had been firmly rejected in the 1940s by Tom Johnston who, as late as 1955, was advising the Labour movement to have nothing to do with such an institution. Others were equally negative. The Royal Commission on Scottish Affairs commented in 1954, for example, that a Highlands and Islands development authority would serve simply to detract from the powers and influence of already existing local authorities. And the Conservative governments which held power between 1951 and 1964 were unrelentingly hostile to the concept of endowing a state agency with the means of intervening directly in economic affairs.

Though post-war Tory Secretaries of State differed markedly on this point from their pre-war predecessors, they were by no means oblivious, it should be stressed, to the difficulties confronting the north of Scotland. The Crofters Commission, after all, was a Conservative creation. A Conservative MP, Lord Malcolm Douglas-Hamilton, played a leading role in the launch in 1953 of the Highland Fund – an independent financial institution which was formed in order to provide the means of supplying crofters and others with low-interest loans for development purposes.

The Highland Fund attracted the enthusiastic support of a Bon-nybridge engineer and businessman, John Rollo, whose intensely-held conviction was that crofting could readily be combined with industrial employment – a belief he had successfully put to the test in the small machine tool plants he had himself established at Easdale, near Oban, and at Inverasdale, in the vicinity of Poolewe. But other prospective employers proved reluctant to follow Rollo's pioneering example. Although the Highland Fund soon acquired a very high reputation among the crofters and fishermen who were the principal beneficiaries of its cash advances, its capital resources – being largely dependent on private donations – were necessarily limited. No such organisation, it was clear, could take the place of government.

This was recognised by Conservative Ministers themselves. 'The Government's Highland policy,' it was stated in a 1959 White Paper, 'is, in general terms, to promote economic growth in the Highlands and to provide suitable amenities and social services so that viable communities with modern standards of living and opportunities for useful employment may be established and retained there.'[1]

It was impossible to take issue with such sentiments; just as it was impossible to maintain that the Highlands and Islands were short of publicly-funded agencies. Indeed one calculation, made by the Highland Panel in the early 1960s, put the total of such bodies at more than 30. None of them, however, seemed capable of taking the concrete measures needed to produce the 'economic growth' to which the Scottish Office was ostensibly committed. There was consequently much frustration of the type encountered by Neil Gunn in the course of his travels with the Taylor Commission:

> One Hebridean crofter showed me correspondence with the county authority, the planning authority, the Department of Agriculture and the Forestry Commission. All were sympathetic, for his agreed scheme of improvements was both intelligent and practical, but these bodies had other disposing bodies beyond them, for cash was involved, not to mention legion interpretations of powers. For example, does the Forestry Commission's remit permit them to grow timber belts on Hebridean islands primarily as shelter belts for stock when with the same expense they could grow more timber elsewhere? Anyway, all that grew out of the crofter's efforts was the correspondence, until he was unable to cope with it and the scheme lapsed.[2]

In the face of such difficulties, and with the depopulation of the north of Scotland continuing as rapidly as ever, there was growing support – in practically every quarter outside the ranks of the Scottish Conservative Party – for the concept of seeking to reverse the region's longstanding decline by means of the formation of a Highlands and Islands development authority of some kind.

The Liberal Party had long supported such an approach. And when, in the general election of 1964, its candidates ousted the Conservatives from several north of Scotland constituencies, this was widely interpreted as an endorsement of the many calls made, in the course of the campaign, for new Highlands and Islands policy initiatives.

But of much more significance – in view of the fact that the 1964 election had provided it with a parliamentary majority – was the position of the Labour Party. Some 11 years earlier Labour's Scottish Council had published a *Programme for the Highlands and Islands* – a programme which advocated the creation of a Highlands and Islands Development Corporation with wide financial and executive powers. That particular proposal had been taken up enthusiastically by the Scottish Trades Union Congress which had organised a series of conferences in Inverness in order to publicise and refine Labour's renewed commitment to the regeneration of the north of Scotland economy. And when, in October 1964, Harold Wilson entered Downing Street at the head of the first Labour government since 1951, the incoming administration's Scottish Secretary, Willie Ross, lost no time in making clear that the establishment of a Highlands and Islands Development Board was his immediate priority.

The necessary Bill was presented to the House of Commons at the start of 1965. The gruffly-spoken Ross, a man not usually given to publicly displayed emotion, introduced the measure to MPs with one of his more impassioned parliamentary performances.

'There is no part of those islands of ours that has merited or received more attention from this House than the Highlands or Islands,' Ross began. Crofters had been given security of tenure. Agricultural and housing grants had been provided. Electricity had been supplied. Roads and bridges had been built. Shipping services had been improved. 'But we must face the fact that we have still the same pattern of unemployment, of underemployment and of a declining population.' Further action was therefore required.[3]

'I think that it has become more and more obvious to everyone who studies the problem,' said the Labour Scottish Secretary, 'that after all the commissions, reports and surveys, and after all the money that has been put into the improvement of agriculture, what really has been needed is an authority with executive powers to deal comprehensively with problems . . . . For this reason we have decided to establish the Highland Development Board.'

When the House of Commons had debated the Crofters Act of 1955, Ross recalled, Labour MPs had warned that the Crofters Commission was not being given the means of making a decisive impact on the High- lands and Islands economy. Time had proved Labour right, the Scottish Secretary now declared. The Commission's powers, 'limited as they were to agriculture', had been shown to be inadequate. And that mistake would not be made again. 'We have to tackle the problem on a wide front and, for that reason, the new Board will require wide powers.'

The HIDB would prepare, concert, assist and undertake a whole series of measures designed to contribute to the development of the Highlands and Islands. It would erect buildings. It would equip and service these buildings. It would establish businesses. It would provide training. It would produce publicity material. It would, above all, disburse generous grants and loans. The overall result, Ross hoped, would be the trans- formation of the north of Scotland's traditionally gloomy prospects.

'For 200 years,' commented the Secretary of State, 'the Highlander has been the man on Scotland's conscience . . . . No part of Scotland has been given a shabbier deal by history . . . . Too often there has only been one way out of his troubles for the person born in the Highlands – emigration.'

And if it were to have the capacity to put past wrongs right, Willie Ross continued, the HIDB would have to deal with the land question. There continued to be 'land hunger' in the Highlands and Islands, and that was unacceptable. 'If there is bitterness in my voice,' said the Scot- tish Secretary, 'I can assure the House that there is bitterness in Scot- land, too, when we recollect the history of these areas . . . . We have nine million acres, where 225,000 people live, and we are short of land. Surely one of the first powers which must be given is a power related to the proper use of the land itself. To my mind, this is basic to any improvement in the Highlands. Anyone who denies the Board powers over land is suggesting that the Board should not function effectively at all . . . . Land is the basic natural resource of the Highlands and any plan for economic and social development would be meaningless if proper use of the land were not a part of it . . . . Clearly, the Board must have power to acquire land, by compulsion if necessary, if it is to be effective.'

Although landowning interests naturally reacted very adversely to the suggestion that the HIDB should have extensive land purchase capabilities, there was, in fact, nothing very novel about such an approach to Highland problems. The Congested Districts Board, formed as long ago as 1897, had been empowered to acquire land for settlement by crofters and, as already mentioned, had done so on a not insubstantial scale. The Board of Agriculture for Scotland, dating from 1912, had also had land purchase powers from the outset and the Land Settlement (Scotland) Act of 1919 had given the Board the right to acquire land compulsorily. By the 1930s, as a result, the state, in the shape of the Department of Agriculture, owned enormous tracts of territory in the Highlands and Islands – that territory having been acquired, for the most part, in order to facilitate the transfer of land from sheep farmers to crofters.

Although land redistribution of the type which thus occurred so extensively in the north of Scotland in the earlier part of the twentieth century is generally assumed to be the prerogative of radical or socialist administrations, most of it was actually brought about by Conservative or Conservative-dominated Governments – Labour, for all its anti-landlord rhetoric, having proved extremely reluctant, when in office, to embark on a worthwhile land reform programme. This was especially true of Clement Attlee's Labour government which refused point blank, for example, to act on the various land purchase recommendations of the Scottish National Parks Survey Committee which had been appointed, towards the end of the war, by Tom Johnston.

The committee, described afterwards by Frank Fraser Darling, who was one of its members, as 'an innocent bunch of idealists', took the view that 'national ownership is part of the natural conception of a national park' and made the point that the whole of the Cairngorms, for instance, could be bought for about £200,000. 'It was extraordinary,' Fraser Darling wrote long afterwards, 'that a forward-looking, experimentally-minded Socialist Government . . . should turn down the idea of national parks in Scotland and that such a Government should have taken some heed of the backwoodsmen and landowners of the opposite side in opposing that decision.' But such had proved to be the case. No Scottish national parks were formed.[4]

Nor was the Attlee administration any more anxious to implement the findings of another of Tom Johnston's creations, the Committee on Land Settlement in Scotland – which recommended that the state should purchase entire crofting townships as a prelude to reorganising them.

Any future land settlement schemes, the Labour government made clear, would be confined largely 'to the creation of full-time holdings which promise a reasonable livelihood'. But nothing was done to put even that comparatively unadventurous suggestion into practice. The Knoydart land raiders, as noted earlier, received no help from the Labour-controlled Scottish Office. The 500 other people who applied to the Department of Agriculture for crofts between 1945 and 1950 were told there were no crofts available. It was left to the succeeding Conservative government to undertake the only effective land settlement project of the post-war period – the establishment of nine new crofts at Craignure in Mull in 1952.[5]

That particular initiative, however, was designed to cater for families who, for reasons similar to those given by the inhabitants of St Kilda 20 years before, had asked the Scottish Office to help them remove themselves from the little island of Soay, off the south coast of Skye. There was not the least indication, throughout the 1940s and 1950s, that Department of Agriculture civil servants had any interest in making use of the sweeping land settlement powers bestowed on them by earlier legislation. That was not at all surprising. Since Department thinking on crofting matters was increasingly characterised by the conviction that existing crofts were agriculturally ineffective, there was little chance of the Department responding sympathetically to pleas for the creation of still more smallholdings.

There ought to be no further disturbance of the sheep farming sector which had already lost a lot of land as a result of previous land settlement, Department of Agriculture representatives told the Taylor Commission. Although Commission members went on to recommend 'that an active and imaginative use should be made of land settlement powers in the crofting counties,' the more worldly-wise of them, at any rate, must have made that suggestion more in hope than expectation.[6]

When, towards the end of the 1950s, the Crofters Commission set up a working party to examine land settlement issues, the working party was obliged to report that 'there was little prospect of land settlement being pursued on any scale at this time'. The possibility of a further land settlement scheme in Mull, the Commission commented, had been gone into in particular detail. But the Department of Agriculture had refused to accept that the piece of ground in question should cease to be farmland. There was an equally negative reaction from the Scottish Office to the suggestion made in 1960 by one member of the Commission, Alasdair MacKenzie, that the government should buy two Skye estates, which were then on the market, with a view to forming a number of new crofts.[7]

But for all the Department of Agriculture's scepticism, the conviction that Highland estates should be broken up as a means of promoting repopulation was one that a significant segment of the Scottish people stubbornly refused to abandon. In 1964 the Highland Panel, which was itself shortly to be replaced by the HIDB, returned to the subject in a report which the Panel's chairman, the leading judge Lord Cameron, considered 'to be as significant and as far-reaching in its implications' as anything which Panel members had previously produced.[8]

'There is a good deal of underused, and in some cases grossly underused, land in the Highlands,' the Panel unambiguously declared. 'Much of this land could be better used. Highland and national interest requires that it should be.' Although 'little or nothing' in the way of land settlement had been accomplished in the north of Scotland since the 1930s, the Panel observed, such land settlement should again be considered. Small crofts of the type created so widely in the past might no longer be appropriate, the Panel conceded. But for entire estates to be devoted to deer stalking was not appropriate either. 'When the primary interest is sport, agriculture suffers,' the Panel declared. The Secretary of State for Scotland should consequently 'exercise his existing powers and acquire suitable land for the creation . . . of family farms'. The resulting farms, in the Panel's opinion, should ideally constitute one element in an integrated package of rural development measures – the other aspects of which might include forestry, tourism and small-scale industry.

In making clear to the House of Commons that the Highlands and Islands Development Board would be expected to deal effectively with the land question, Willie Ross had made specific reference to the Highland Panel's 1964 report. That report had also generated a good deal of interest in the north of Scotland. The HIDB's first chairman, the eminent Scottish planner Professor Robert Grieve, was consequently under no illusions as to the importance which both the public and the politicians were likely to attach to the new agency's activities in the more rural parts of the Highlands and Islands.

'No matter what success is achieved in the Eastern or Central Highlands,' Grieve noted in a paper he prepared in March 1966, 'the Board will be judged by its ability to hold population in the true crofting areas.'[9] In the opinion of the Highland Fund's John Rollo, appointed HIDB deputy chairman in recognition of his longstanding involvement in the Highland development effort, that implied the widest possible dispersal of industrial employment.

'There is a tendency in certain planning minds to think in terms of concentrations of population around existing towns or large centres . . . with a view to factory development,' Rollo observed in an article he contributed to *The Scotsman* in February 1965, some nine months before the HIDB's inauguration. 'This will be fatal for the Highlands because it will result in the complete destruction of the agricultural crofting population. Far more effective would be concentration on the fostering and development of small factories located where the people are and have their homes and so allow them to have wage-earning work and work their crofts as an ancillary to give increased income.'[10]

But for all that the HIDB acknowledged of crofting that 'if one had to look now for a way of life which would keep that number of people in such relatively intractable territory, it would be difficult to contrive a better system,' neither Professor Grieve nor his senior staff gave the impression that they shared Rollo's views. Rollo himself occupied an increasingly marginal position in the Highland Board's Inverness headquarters. It was perhaps indicative of the HIDB's priorities – or, possibly, of the restrictions placed on its operations by the Department of Agriculture to which the Board was initially answerable – that by the end of 1966, when the agency had been in existence for 14 months, not one land use specialist had been appointed to its staff.[11]

The Western Isles MP, Malcolm MacMillan, ever suspicious of officialdom's attitude to crofting, had warned in the course of the parliamentary debates on the Highlands and Islands Development Bill that it was essential to 'stabilise population' in the more rural parts of the north of Scotland. Like Rollo, MacMillan feared that there would be undue concentration on one or two essentially urban localities. 'I hope that the Board will not be thinking in terms of the traditional idea of growth points,' he said.[12]

But this was exactly what Grieve had in mind. 'From the beginning,' as one of the Highland Board's early reports remarks explicitly, 'we decided to concentrate attention on areas that have special advantages and resources for growth.'[13]

This 'growth centre' strategy was destined to end in tears. By the 1980s and 1990s, the closure of the Fort William pulp mill, together with the demise of the Invergordon aluminium smelter and the steady run-down of the Dounreay nuclear reactor establishment, had transformed Lochaber, Easter Ross and Caithness – which the HIDB had originally identified as its three best prospects – into 'problem areas' characterised principally by persistently high rates of unemployment. But these future difficulties, quite understandably, were not generally foreseen in the mid-1960s. 'It would be daft,' one HIDB board member is said to have remarked at that heady time, 'to try and catch sprats when they could catch a whale.' Amid all the confident talk about the need to establish something called a 'linear city' on the east coast littoral between Tain and Nairn, John Rollo's vision of a reinvigorated crofting economy tended, like its author, to be lost to sight.[14]

'We are belaboured equally,' Professor Grieve commented ruefully, 'by those who say we should be taking over great areas of land and splitting them up into small farms; and by those who regard the crofting way of life as false, uneconomic, rotted by subsidies and leading to an inevitable, if euphoric, death.'[15] But there was little doubt in which direction the Highland Board was tending.

Agriculture, the HIDB acknowledged, would continue to be 'an important part of the Highland economy'. Certain crofting and farming activities might even be capable of expansion. It was for this reason that the Board was of the opinion that 'the wider availability of capital would be the most effective stimulus to Highlands and Islands agriculture'. HIDB aid to the agricultural sector would consequently have 'a bias towards capital-intensive rather than labour-intensive development'. It was 'very clear', therefore, that any growth in agricultural output was 'highly unlikely to be accompanied by an increasing population . . .

rather it is likely to be accompanied by a drop because of more efficient and more mechanised methods.' And so the overall result of any agricultural improvement effort would be 'more food from the Highlands for the rest of the country rather than more people for the Highlands'.[16]

This, of course, was broadly to subscribe to the gospel which the Department of Agriculture had been preaching since the 1940s. In only one – initially very limited – respect did the new development agency evince any interest in embarking on genuinely novel experiments of a type which might actually result in radical changes in land use. Many of the north of Scotland's more isolated localities, it was noted briefly in the Highland Board's first annual report, 'could sustain two or more different kinds of activity – agriculture, afforestation, tourist accommodation, recreation and sport, small industry – with great benefit to their economic and social life'. The HIDB accordingly proposed to inquire closely into the possibility of mounting two 'comprehensive development' projects of this kind – in the island of Mull and the Strath of Kildonan.[17]

There was no very obvious connection between 'comprehensive development' – which implied the continuance of the highly diversified rural economy traditionally associated with crofting – and the HIDB's much more explicitly stated commitment to fostering the emergence of more specialised forms of agriculture. It is by no means implausible, therefore, that it was, as one member of its staff wrote subsequently, 'with some reluctance', and only under pressure from Labour ministers, that the HIDB agreed to set foot on the comprehensive development road – such claims being lent considerable credence by the tenor of the Board's report on its findings in the Strath of Kildonan.[18]

That particular part of the Highlands had been one of the places most depopulated as a result of the mass evictions conducted by the Sutherland estate in the early nineteenth century. Because of its close association with the 'history of the clearances', coupled with continuing 'absentee landownership and apparent underuse of land resources', the Strath of Kildonan, in the HIDB's opinion, had 'been frequently cited as a typical example of misused land'. Its development 'along comprehensive lines' had accordingly been suggested by the Highland Panel in 1964. It was for those 'historical' reasons – 'and not because of any particular known resource or combination of resources' – that the Highland Board had 'decided to examine the Strath's potential'.[19]

That potential, the HIDB reported in 1970, was generally limited. Some 15,000 acres might be usefully afforested. A manufacturing or processing venture might be established in the coastal village of Helmsdale – where the fishing fleet might also be expanded. And there was scope for further tourism. But no case could be made 'on economic or employment grounds' for land settlement. Such settlement, the Board pointed out, would be 'contrary to the spirit' of the Agriculture Act of 1967 which, as noted in a previous chapter, was intended to promote farm amalgamations. The HIDB had 'no desire to set Highland farming on the reverse process.'[20]

These conclusions, as the Highland Board was well aware, were likely to lead to charges of timidity on its part. The agency was consequently anxious to rebut any notion that it was unduly susceptible to pressure from landowners. 'If the Board believed that the establishment of smallholdings in the Strath of Kildonan would be beneficial,' the HIDB commented by way of explaining its negative attitude to land settlement, 'it would not hesitate to recommend that the Secretary of State take the steps needed to bring about such a land reform . . . . If land settlement in the Strath of Kildonan had been considered a practical proposition, any unreasonable objection to such a programme by the landowners would not have prevented the Board from recommending such a settlement policy.'

But if Sutherland offered no land settlement possibilities, what of Mull – an island which, according to Frank Fraser Darling's *West Highland Survey*, contained 'some of the most fertile land' in the north of Scotland? Mull, Fraser Darling had observed, was a classic example of 'a countryside ruined by . . . the sudden introduction . . . of sheep farming on the extensive system'. Many parts of the island, he told the Taylor Commission in 1952, should immediately be given over to crofters by means of a major land settlement programme of the kind put into practice in Skye in the 1920s.[21]

Fraser Darling's prescriptions for Mull, needless to say, had not been acted upon. With its population continuing to spiral steadily downwards through the 1950s, the island was widely considered to exemplify those processes of decline from which so many parts of the Highlands and Islands were then suffering.

'We have given much thought to the problems of Mull where severe depopulation has taken place and where the standard of agriculture has suffered from a marked trend towards the grouping of farms and the merging of farms into sporting estates,' commented the Highland Panel in 1964. 'It has been clear for some time that, unless something radical is done, the island of Mull will have declined to the point at which the revival of community life and a viable economy might no longer be possible.'[22]

If the HIDB was seriously to promote land settlement, then, Mull was evidently the place to start. And in Mull, the HIDB confirmed in 1968, it was indeed investigating 'the possibility of creating a number of commercial farms from land surplus to forestry or from land purchased from large estates'.[23] The results of these investigations were eagerly awaited.

Again, however, there was disappointment. The Mull report, admittedly, was a good deal more positive than its Kildonan predecessor – going so far, for instance, as to maintain that a number of new farm tenancies might be created and advocating an extensive land improvement programme. But what was becoming increasingly apparent, to the Highland Board itself as well as to its growing number of external critics, was that it was one thing to produce a well-researched report of the Mull or Kildonan type; it was quite another to put that report into effect.

'The problems associated with drawing up detailed land use proposals have been, at least in retrospect, minor compared with the problems of implementation,' the HIDB complained in the mid-1970s. Some progress could be made on land already in the ownership of other public sector organisations such as the Forestry Commission and the Department of Agriculture. But prospects for change on privately-owned estates were much more problematic. 'The attitude and motivation of those who currently own the land,' the Board commented, inevitably constituted 'an important influence' on development – 'and while many farmers and others were concerned to improve the land . . . some estates were acquired only for residential and sporting purposes and land improvement was not an important part of estate policy.'[24]

In both Mull and the Strath of Kildonan, the HIDB conceded publicly, there were 'differences in opinion and objectives' between itself and a number of landowners. 'What seems to be lacking,' the HIDB continued, 'is a satisfactory means of resolving these conflicts which will give due weight to the regional development priorities of the Board and, implicitly, to the needs of the community.'

A reading of the parliamentary debates surrounding the passage of its founding Act, the HIDB contended, accurately enough, suggested that 'the powers given to the Board were intended to provide the means for the Board to implement its land development proposals should these be frustrated by landowners'. But these powers, as the HIDB was now learning from experience, were turning out to be incapable of any such application.

It was not impossible for the HIDB to acquire land. In 1970, for instance, the Board bought a substantial slice of the Cairngorms from the Forestry Commission. And in 1977 it was to purchase the Rahoy Estate in Morvern in order to establish an experimental deer farm. What proved much more difficult, however, was the acquisition of an estate which was large and varied enough to allow the HIDB to demonstrate the full range of land use possibilities – as outlined, for example, in its Mull report.

In 1975, in a further attempt to make good this shortcoming, the HIDB decided to buy the 9,500-acre Killiechronan Estate in Mull – only to be told that the Treasury would not agree the necessary expenditure. Although bids were also made for other properties – including the island of Eigg – these proved, for various reasons, equally unsuccessful.

The exercise of the HIDB's compulsory purchase powers was considered on two occasions. But for all that Willie Ross in 1965 had repeatedly stressed the 'comprehensive' and 'sweeping' nature of the powers at the Board's disposal, Board members were now legally advised, somewhat to their surprise, that these powers were 'not appropriate' for the acquisition of the properties they had in mind. The purchase powers with which the HIDB had been equipped, it had belatedly appeared, were essentially identical to those available to local authorities. They were designed, therefore, to facilitate the procurement of a relatively limited area for the improvement of a road or the construction of a building. They were not intended to enable a public body to take over an extensive tract of land for largely agricultural purposes. For this reason, it was announced in the mid-1970s, the HIDB had decided to seek an enhancement of its powers in order to enable it to deal effectively with 'obvious examples of underuse or mismanagement of land which are hindering the development of rural communities or even endangering their future existence'.

Public discussion of Highland land use issues often generated more heat than light, HIDB chairman Sir Andrew Gilchrist commented. 'However, if a good deal of what is said on land use is somewhat ill-informed, that does not mean that there is no underlying problem. There is a problem; there is very considerable underuse of land in the Highlands and Islands. I think that the Board . . . may have been slow in coming to grips with this situation and, to that extent, we must expect criticism. All I can say is that it is a very difficult problem with strong political implications.'[25]

Gilchrist, a former diplomat who had been appointed HIDB chairman by the Conservative government following Ted Heath's electoral victory in 1970, had begun by describing the Board as 'a merchant bank with a social purpose'. This did not seem to presage any very radical new departures on the land use front. Nor did Gilchrist's conviction that it was wholly unfortunate that the impression had ever been given that the Highland Board was formed with the purpose of tackling the landownership issue. With its relatively limited financial resources, he commented retrospectively, the HIDB could not possibly ever acquire more than one or two estates. 'Thus in the doubtful hope of creating . . . a few jobs . . . the Board would find itself at the storm centre of not merely a legal but a political struggle which ought to have been fought out in Parliament.'[26]

But for all his suspicions – which, in the end, were to prove well-founded – that any HIDB venture into the ownership minefield was likely to prove abortive, Gilchrist was persuaded to launch the Board's first serious attempt to utilise the purchase option as a means of removing an evident obstacle to development.

His target was one of the more notorious landlords in the north of Scotland – Dr John Green, a Sussex medical practitioner who had first bought property in Raasay in 1961 and who had subsequently gone out of his way to frustrate successive local authority attempts to provide Raasay with an urgently needed ferry terminal.

'I was well aware of the legal resistance which might be made,' the HIDB chairman afterwards recalled of the Board's plan to buy out Green and to provide Raasay with the means of advancing economically. 'But I relied on the conspicuous contrast between our own development plans and stagnation and dereliction under the existing ownership. Surely we could defeat a bad landlord and take over his land by exhibiting our legal powers and by proving we would be a better landlord.'[27]

It was in relation to Raasay, however, that the Highland Board first learned that its land acquisition capabilities were not nearly so extensive as had been previously assumed. Green was to remain stubbornly in possession until 1979. An increasingly aggravated Gilchrist meanwhile authorised John Bryden, the young agricultural economist who had been put in charge of the HIDB's Land Division, to undertake the policy review which was to result in the Board seeking to augment its purchase powers.

The later 1970s were a period of mounting public interest in, and concern about, the nature of Highlands and Islands landownership. The extent to which such ownership remained the prerogative of a tiny number of people was convincingly demonstrated by the 90-year-old, but unrelentingly radical, John MacEwen in a book entitled *Who Owns Scotland?* The resulting debate as to the rights and wrongs of this situation was given added urgency by media coverage of the activities of the Dutch, Arab and other overseas purchasers who were increasingly active on the north of Scotland land market.

'Nearly all the estates sold in the Highlands in the last three years have gone to foreigners,' the Scottish Landowners Federation reported in 1977. 'This is a significant trend,' the SLF conceded, 'and would be quite unacceptable in the long-term if continued.'[28]

That some of the north of Scotland's new lairds were not without a sense of humour was evident from the names of their landholding companies – not the least picturesque of these being Perfidia Investments which acquired 10,600 acres in Glen Dessary in 1977. But the fact that such concerns were habitually registered in places such as Panama, the Channel Islands, Lichtenstein and Gibraltar – where the identity of their owners could be kept forever confidential – led to increasingly adverse comparisons being made between the Scottish position and that prevailing in most other European countries where land dealings of so secretive a sort were legally impossible.

There were repeated demands from the Scottish National Party, then enjoying an unprecedented level of electoral support, for Scotland to be protected from further land speculation by legislation – of a type common on the Continent – designed to limit ownership to UK nationals. And in 1977 James Callaghan's Labour government agreed to appoint a committee, chaired by Lord Northfield, to inquire into the overall implications of changes in the ownership of Britain's countryside.

In relation to these wider controversies, the HIDB's *Proposals for Changes in the Highlands and Islands Development (Scotland) Act to Allow more Effective Powers over Rural Land Use*, as revealed in June 1978 by Professor Kenneth Alexander, the Strathclyde University economist who had become Highland Board chairman in succession to Sir Andrew Gilchrist, seemed almost excessively modest.

> The Highland Panel bequeathed to the Board in 1965 the task of investigating the Strath of Kildonan and Mull. The Board published reports and recommendations on both areas, following extensive survey and consultation work, and attempted to persuade key people to implement the recommendations. Years later, only small advances can be seen in respect of land in private ownership. In both areas, a relatively small number of landowners is involved in key developments, but the Board is not empowered under present legislation to push through the crucial measures.[29]

This experience, the HIDB *Proposals* continued, was indicative of 'the clash which can, and does, occur between the interests of owner and community in the Highlands and Islands and emphasises the measure of power, for good or ill, which can be wrought by one private individual over the lives of whole communities'. And it was this ability of estate owners to act counter to the community interest, the Highland Board made clear, that it was now attempting to curtail.

Were its proposals to be implemented, the HIDB explained, particular localities – such as the Strath of Kildonan – might, in future, be designated as areas in which the Board would be entitled to exercise the new powers it was now formally seeking from government.

The designation of any given area would normally be initiated by the HIDB itself. But local authorities, farming organisations and, for that matter, any group representing at least 10 per cent of the registered voters resident in the area would also be entitled to start designation procedures.

Once the Board had accepted the case for designation, a local advisory committee and a technical panel would be constituted. The first would consist of nominees of the local authorities, farming and landowning groups, the Scottish Trades Union Congress and the HIDB itself; the second would be representative of the Agricultural

Colleges, the Department of Agriculture, the Hill Farming Research Organisation, the Royal Institute of Chartered Surveyors, the Institute of Foresters of Great Britain and the Crofters Commission.

In consultation with both committees, the Highland Board would prepare a 'draft outline development plan' for the area. This plan would be sent to all affected owners and occupiers of land, and discussions would be opened with those owners and occupiers. The local advisory committee would hear objections from any landholder wishing to take issue with the plan. The committee, in collaboration with the technical panel, would then make its views on the plan known to the Board which, in turn, would make any appropriate amendments prior to publicising the revised plan in the designated area. Finally, the Board, in collaboration with the local advisory panel, would organise a public meeting at which all interested parties would have the opportunity to make still more comments in advance of the Board submitting the finalised plan to the Scottish Secretary for his approval.

Should the necessary ministerial consent be forthcoming, the HIDB would subsequently be able to deploy – in the designated area – a number of additional powers. These would include the power of nominating tenants to such land as the Board might schedule for leasing on private estates; the power of controlling land sales in order to ensure that such sales did not result in the Board's developmental intentions being obstructed; and the power of compulsory purchase which, the Board hastened to stress, would be used only as an 'instrument of last resort'.

The HIDB's land reform proposals, then, were rather less than revolutionary. As was shrewdly noted by Brian Wilson of the *West Highland Free Press*, which inevitably took an especially close interest in such matters, this was to be expected. 'An initiative by a public Board, limited by Statute,' Wilson wrote, 'can never be a substitute for a real political initiative which only a government can undertake. Until a Labour government adopts a radical land policy which does not aspire to neutrality in the irreconcilable conflict between landlordism and the good health of society, then attempts at ameliorative action – however well intentioned – are bound to be disappointingly and frustratingly inadequate.'[30]

Had the Highland Board actually obtained the powers it sought, very little would have changed. The complex designation procedures outlined in its 1978 paper bore more than a passing resemblance to those which the Crofters Commission had been obliged to follow in the course of its attempts in the later 1950s to reorganise crofting townships like Big Sand. Although the HIDB gained considerable public support for its proposals – including, it should be stressed, the backing of one or two of the north of Scotland's more progressive landlords – it is highly probable that, had it been given the legal authority to put its thinking into practice, the Board, like the Commission before it, would have become hopelessly enmeshed in virtually endless rounds of consultation and dispute.

In the event, the Highland Board's resolve was never tested. Its request for enhanced powers was formally made to Bruce Millan, Callaghan's Scottish Secretary. The Prime Minister, while on a visit to the north of Scotland, had himself hinted at some sympathy with the HIDB's case, but Millan was evidently in no hurry to respond. His attitude was still unclear when the Labour government fell in the spring of 1979.

It was left to Millan's Conservative successor, George Younger, to administer the *coup de grâce* to what remained of any hopes that the HIDB might prove itself a vehicle of land reform. Asked in the House of Commons by Caithness and Sutherland MP Robert Maclennan if he would 'seek to increase the powers of the Highlands and Islands Development Board over the use of land', Younger replied: 'No, sir. The HIDB has similar powers to those of other public agencies and local authorities and I am not persuaded that there is need to extend them.'[31]

That the Highland Board proved incapable of fulfilling the more radical expectations aroused by Willie Ross in 1965 was arguably more the fault of politicians than of the HIDB itself. It was fortunate for the crofting population, therefore, that the Board's failure to gain Scottish Office backing for its land reform plans did not weaken its growing determination – a determination resulting partially from the manifest failure of its earlier 'growth centre' strategy – to promote a wide range of developmental initiatives in the more rural parts of the Highlands and Islands. It was equally fortunate, in view of the stringent public spending policies which were to be the hallmark of Margaret Thatcher's Conservative administrations throughout the 1980s, that the HIDB was not left to carry this burden unaided.

When, in 1975, the Labour government gave the British electorate the opportunity of voting in a referendum on whether or not the country should remain a member of the European Community, which the UK had joined some two years earlier, opposition to the EC link was strongest in peripheral localities like Shetland and the Western Isles – where there was understandable apprehension that the process of European integration would result in still fewer economic opportunities being made available to those regions most remote from the EC's Brussels-centred heartland. But it is arguable, in retrospect, that the islands were lucky to have been outvoted by the rest of Britain on the European issue.

In 1979, when the HIDB was still endeavouring to win political approval for its land reform package, rumours began to circulate that the European Commission was contemplating a substantial developmental venture in the north of Scotland. These rumours proved correct. In 1982 the EC launched a five-year Integrated Development Programme in Lewis, Harris, the Uists and Barra. This Western Isles IDP was followed, in the remainder of the Hebrides and in the Northern Isles, by an Agricultural Development Programme. And in 1991, while the ADP was still in progress, four localities on the Highland mainland – Lochaber, Lochalsh, Wester Ross and the eastern part of Sutherland – became the first recipients of yet more EC aid in the shape of a Rural Enterprise Programme.

The general thrust of those various measures – which together injected almost £100 million into the north of Scotland countryside – was primarily a matter for the European Commission. But all three programmes were financed jointly by Brussels and Whitehall. Their impact was enhanced, as UK government ministers understandably stressed, by the more modest – but by no means negligible – endeavours of the HIDB.

Impressed by the extent to which Western Isles crofters had capitalised on the opportunities offered by the IDP, which provided grant aid for a whole range of crofting improvements, the Highland Board embarked on its own Skye Development Programme in 1986. A North West Development Programme, focussing mainly on the more northerly and westerly parts of Sutherland, followed two years later. Among the many other crofting-orientated projects in which the HIDB became involved as the 1980s advanced were a series of both local and regional stock marketing, machinery sharing and animal health initiatives – all of them intended to add to crofting incomes.

To these various programmes there was an immediately enthusiastic response from crofting communities. This was evidently a matter of some surprise to those still subscribing to the theory that the crofter was an irredeemably sluggish individual. In fact, however, it had long ago been proved, most particularly by two remarkably energetic men, Roddy MacFarquhar and Archie Gillespie, that crofters were perfectly capable – given appropriate encouragement – of demonstrating those qualities of enterprise and vigour which their many external critics had long claimed they lacked.

MacFarquhar, who had fought with the Communist-led International Brigade in the Spanish Civil War and who subsequently contributed substantially to the policy rethink which resulted in the Labour Party committing itself to the establishment of a Highlands and Islands development agency, became involved in crofting matters as a result of his being employed first by the Scottish Agricultural Organisation Society and later by the Highland Fund. While serving as an SAOS field officer in Uist in the 1950s, MacFarquhar founded the first commercially successful crofting co-operative – to handle the marketing of eggs.

Other equally well-supported ventures were soon to follow – with the agricultural supplies co-operative Lewis Crofters, for example, quickly acquiring more than 2,000 crofting shareholders on its launch in 1958. This was impressive enough. But it was what followed the arrival in Lewis at that time of the Islay-born North of Scotland College of Agriculture adviser, Archie Gillespie, which really showed that the crofting community remained capable of generating tremendous effort on its own behalf.

Gillespie's achievement was to appreciate the full potential, in Hebridean conditions, of enhancing the fodder-producing qualities of peaty moorland by means of applying to it compound fertiliser, a grass-clover seed mix and, above all, huge quantities of lime-rich shell sand of a sort that was readily available on island beaches. Surface reseeding of this type, Gillespie believed, could greatly enhance island livestock production, both in quantity and quality. By dint of much exhortation, not to say personal example, he persuaded scores of Lewis and Harris township committees to embark on extensive reseeding schemes.

'Everyone thought he was crazy,' an elderly Lewis crofter subsequently recalled of Archie Gillespie's first visit to one particular community. But that crofter and his neighbours were nevertheless impressed by the agricultural adviser's sheer fervour for reseeding, and they eventually agreed to put a project of their own in hand. 'The whole village turned out to put down the sand,' the same crofter remembered. 'We did it by hand, a string of about 30 of us with buckets. We did the same with the seed and the fertiliser.' Soon thousands of acres were being given identical treatment.[32]

'It is quite heartening,' remarked Western Isles MP Malcolm MacMillan in 1961, 'to go into one area after another on the island of Lewis and see on one side of the road the brown, barren moorland, which nobody had ever thought could be contemplated for regeneration and improvement, and, on the other side of the road, these great stretches of beautiful green feed, making such a difference to the area.'[33]

The Scottish farming press – which had not previously been noted for its appreciation of the crofting community's agricultural abilities – was, if anything, even more complimentary. But it was left to Gillespie himself to highlight the really decisive transformation produced by his reseeding programme. 'The change in the landscape is a great thing,' Archie Gillespie said in 1963. 'But to us who are in close contact with the job the very big thing is the improved mental climate.'[34]

Crofters grown accustomed to the notion that they were the victims of unalterable circumstances which were necessarily condemning their communities to perpetual stagnation and decline, in other words, had learned that it need not be so; that there were steps that they themselves could take to improve their own prospects.

Subsequent experience was to reinforce that lesson. Although Catriona MacLean and George Campbell, the young and immensely energetic project officers whom the HIDB employed to give effect to its development programmes in Skye and Sutherland in the later 1980s had not even been born when Archie Gillespie began his work in Lewis, they stood very much in the Gillespie tradition. Like their immediate superior Sandy Cumming, who then headed the Highland Board department responsible for the Skye and Sutherland initiatives, MacLean and Campbell believed passionately in the worth of crofting. To be possessed of such conviction in relation to a landholding system which so much of officialdom had been disparaging for decades was to display qualities which most representatives of public bodies had for far too long been lacking.

The crofting resurgence which men like MacFarquhar and Gillespie had helped to originate, and to which the IDP, the ADP and associated undertakings contributed so substantially during the 1980s, had first begun to become generally apparent in the course of the 1970s. It was then, for instance, that Highland Board chairman Kenneth Alexander – whose lead, in this instance, was to be broadly followed by his successor, Robert Cowan – started to place a wholly new emphasis on the possibility of engendering development activity in those crofting localities which the Board had initially been inclined to leave largely to their own devices.

HIDB staff had previously been based entirely at the Board's Inverness headquarters. Now offices were opened in places like Lerwick, Stornoway, Benbecula and Portree. Substantial financial assistance was provided to crofter-controlled marketing groups – such as Skye Livestock, Lewis Livestock and Uist Calf Producers. There was heavy investment in fish farming – which, by the end of the 1980s, was employing some 1,600 people on the west coast and in the islands. Although there was considerable criticism of the extent to which fish farming was dominated by very large commercial concerns, over which most people living in the Highlands and Islands could exercise little influence, let alone control, this criticism was countered, to some extent at least, by one of Alexander's more imaginative initiatives – the injection of HIDB funds into community co-operatives.

Ventures of that type had been pioneered in the west of Ireland as a means of mobilising both the financial resources and the human talents which are invariably to be found in even the most remote locations. The Highlands and Islands community co-operatives which the Board began to help in the later 1970s were accordingly structured, like their Irish counterparts, in such a way as to enable them to involve considerable numbers of rural residents in the management of a wide range of business enterprises.

'I believe that the value of such community activity lies as much in what it can do for morale and self-confidence as in its direct economic impact,' Alexander commented in 1978. In this sense, in his opinion, HIDB backing for the co-operatives then being established in so many crofting areas constituted a challenge to the 'ingrained' tendency, on the part of people in the Highlands and Islands, 'of looking to the outside for assistance'. Greater self-reliance had to be encouraged. This was what community co-operation was all about. 'A collective or community approach to development can foster self-confidence and a spirit of independence within the community by demonstrating that progress can be made under local leadership and need not lean so heavily upon outside influence and help.'[35]

What one commentator called a 'new mood' could indeed be detected now in many crofting areas.[36] Depopulation was slowing, or being halted, very widely. In some places – most notably Shetland, which was deriving substantial financial benefits from the arrival in the islands of the offshore oil industry – substantial repopulation was at long last getting underway.

Shetland Islands Council was channelling a significant proportion of the revenues accruing from its various deals with the oil companies into projects designed to stimulate the local crofting economy. And Comhairle nan Eilean or the Western Isles Island Council – one of the more positive results of the 1975 reform of Scottish local government in that the Western Isles, previously administered from the faraway towns of Inverness and Dingwall, now had their own local authority – was making highly imaginative use of the government-funded Job Creation Programme in order to combat the unemployment from which the Outer Hebrides had suffered for so long.

'Job Creation,' it was noted at the time, 'had been a rare example of a national scheme which proved to be specially suitable for conditions in the crofting areas.' Township committees were closely involved in the Comhairle's job creation effort from the outset. The overall rate of unemployment in the islands was more than halved. Valuable training was provided. Crofting communities were equipped with badly needed fanks, peat roads, jetties and other facilities.[37]

Wester Ross crofters were employed at the oil production platform fabrication yard at Kishorn. Lewis crofters worked at the fabrication yard at Arnish. Easter Ross and Sutherland crofters got jobs in the various oil-related enterprises which had mushroomed in the vicinity of the Cromarty Firth. Nor was the overall improvement in crofting conditions simply a matter of there being more work – and more money – available. Throughout the Highlands and Islands, it was clear,

171

crofting communities were beginning to take more pride in their cultural heritage, their history, their identity.

Books and articles about the accomplishments of the nineteenth century Land League began to be published. A play by John McGrath, *The Cheviot, the Stag and the Black, Black Oil*, was taken on tour to countless Highlands and Islands village halls – where capacity audiences clearly warmed to its message that the time had come for people in the north of Scotland to stand up against the various commercial interests which, in McGrath's opinion, had been ruthlessly exploiting them for generations.

Gaelic music was flourishing as never before. With the emergence of a number of Gaelic rock bands, of which the most successful was Runrig, made up of a number of lads from Skye and Uist, there were heartening indications that the language need not be eternally associated with older people and their nostalgic yearnings for a vanished past.

There was growing interest, too, in the dialect and music of Shetland. Alongside this renewed emphasis on what it was that distinguished Shetland from the rest of the United Kingdom, there emerged something approximating to a nascent Shetland nationalism – with the Shetland Movement, supported by a significant number of local councillors, demanding that Shetland be given political autonomy of the type enjoyed by that other North Atlantic island group, Faroe, since the 1940s.

Faroe – which, by Highlands and Islands standards, was strikingly prosperous as a result of its success in declaring and defending an exclusive fishing zone of the sort denied by British governments to Scotland's island and coastal communities – also attracted the attention of that most untypical crofting landlord, Iain Noble. A merchant banker and Gaelic enthusiast who had bought extensive properties in Skye, Noble was well aware that the regeneration of Faroe had commenced with the revival of its language.

The pride which had begun to be taken in spoken and written Faroese in the 1930s, Noble asserted, was the key to that enhanced self-esteem which had led both to Faroese self-government and to the massive expansion of the Faroese economy by means of the establishment of locally controlled, locally operated enterprises of a kind that were conspicuously lacking in most of the north of Scotland. At the start of the twentieth century, Noble pointed out, the population of both Skye and

Faroe had been roughly equal – at around 15,000. But whereas Skye's population had subsequently fallen by almost half, Faroe's had increased by a factor of three. With the explicit intention of helping to stimulate a similar renaissance in the Highlands and Islands, Noble founded a Gaelic-medium college at Sabhal Mor Ostaig in Sleat.

That college – located, symbolically enough, in the refurbished steadings of one of the many large farms created in the course of those nineteenth-century clearances which came so close to eradicating both crofting and Gaelic – was to expand steadily in the 1980s. And so was Gaelic-medium education generally – the Gaelic-medium playgroups formed in the 1970s being followed eventually by the provision of more and more Gaelic-medium classes in Highlands and Islands schools.

BBC Radio nan Eilean began broadcasting in Gaelic from Stornoway. A Gaelic drama company was formed. Ambitious community education projects were being launched in crofting areas. Local history societies were everywhere becoming active. A Gaelic publishing company was founded. Wherever one looked in the north of Scotland, it seemed, there were welcome indications of vitality of a kind not seen in much of the Highlands and Islands for very many years.

But in all of this, observed Brian Wilson of the *West Highland Free Press*, which vigorously championed and promoted developments as diverse as community co-operatives, livestock marketing groups and Gaelic-medium schools, one element was conspicuous only by its absence.

'Crofters . . . appear to have lost their collective voice,' Wilson commented. The Crofters Unions which had been so active in the 1960s, he continued, were now 'pretty moribund organisations'. But for all the apparent failure of what had gone before, there remained a pressing need 'for a strong and articulate organisation to keep the crofting case to the fore'. It was essential, in Wilson's view, that such an organisation be rapidly created.[38]

> The time is ripe for the Crofters Unions to be given a new lease of life and to become meaningful pressure groups harassing central and local government. Those who have tried in the past to stimulate such interest will doubtless point out that all this is easier said than done. Those who are most actively engaged in crofting do not, by definition, have time for the meetings and administrative functions that close interest in a lively Crofters Union would demand. There are also major problems of communication.

The answer could lie in the appointment of a full-time union organiser and spokesman. Would the HIDB perhaps put up the money initially to finance such a post? Certainly it is time that local authorities made some small acknowledgment of the fact that crofting agriculture is the economic basis – the *sine qua non* – of their whole area. They should chip in. On a whole range of issues, crofters . . . need someone to speak up for them. Bold initiatives are urgently required.

Chapter Seven

A VOICE FOR CROFTERS

Brian Wilson's article advocating the reactivation of the Crofters Union movement appeared in October 1979. The following month, Charles MacLeod, the former chairman of the Federation of Crofters Unions and a man with whom Wilson had clashed frequently on the owner-occupation issue in the period preceding the Crofting Reform Act of 1976, addressed the Crofters Commission's assessors conference in Inverness.

The Commission had got into the habit of referring to this annual gathering as 'a crofter parliament', MacLeod remarked. But this, he said, 'was surely stretching exaggeration quite a bit'. Though the function of the assessors was 'to feed the Commission with grass-roots opinion', they were, in essence, Commission appointees. 'The assessors panel never meets independently of the Crofters Commission,' MacLeod pointed out. 'It does not have a chairman of its own or other office-bearers, self-chosen.' There was no sense, therefore, in which the assessors conference could be regarded as a substitute for an effective union.[1]

'A Crofters Union,' MacLeod stressed, 'meets independently. It can kick things about independently and present a corporate voice to the Commission and other bodies . . . . Not being related by quango blood-ties to the Commission, as the assessors panel is, a union can berate or criticise the Commission with impunity, if such is warranted.'

The chain of events leading to the formation of the Federation of Crofters Unions in the early 1960s, Charles MacLeod recalled, had commenced with the Crofters Commission going 'off their heads' and advocating a massive reduction in the number of crofts. The Federation had done 'some useful work' – not least in ensuring that the Commission's amalgamation proposals had never been put into practice. But the Federation, MacLeod acknowledged, was now 'moribund', and a fresh start was required.

175

Such a new beginning was all the more essential, though Charles MacLeod, quite understandably, did not say so, as a result of the extent to which the Federation of Crofters Unions had become identified, in the early 1970s, with the view that crofters should become the owner-occupiers of their holdings. By 1976, when the legislation making possible a general transition to owner-occupancy was finally enacted, both the Federation and the Lewis and Harris Crofters Union – the two being by that point more or less identical – were being openly condemned for their failure, as many crofters saw it, to keep properly in touch with crofting opinion.

The Lewis and Harris Crofters Union, said one Point crofter in 1976, was a 'most unrepresentative body'. Two members of its executive, the same crofter pointed out, were also 'salaried members of the Crofters Commission – a very disturbing situation'. Still more disturbing, perhaps, was the mounting evidence that crofters were no longer paying union dues. This, one crofter alleged at the following year's annual general meeting, was partly owing to the fact that the Crofters Union was identified too closely with the Western Isles Constituency Labour Party as well as with the Crofters Commission. 'The idea that the Crofters Union is allied to any political party died a natural death years ago,' one loyal union man insisted. Said a more sceptical member of the audience: 'So did most of the membership.'[2]

And that, unfortunately, was all too obvious. Even in Lewis and Harris, long the principal centre of the Crofters Union movement, no more than 300 or 400 subscriptions were being collected annually. Total membership elsewhere in the Highlands and Islands did not approach even that very modest figure. In many places – such as Benbecula, South Uist, Barra, Skye and several parts of the mainland – formerly active unions had ceased effectively to function.

It was about this time, however, that the man who was to contribute most to reversing this decline began to take an interest in union affairs. Angus Macleod, a Lewisman who was then well into his sixties, came originally from the crofting township of Colbost. One of the few small-scale island tweed producers to have survived independently of the major mills in Stornoway, Macleod had also been a retailer and a guest-house proprietor. Although actually one of those rare individuals whose politics become steadily more radical with age, his entrepreneurial activities had resulted in his being regarded – certainly by most of the

Labour activists who had earlier been to the fore in Crofters Union circles – as a Tory. This is not a very popular thing to be in the Western Isles where Conservative candidates in general elections have tradition-ally come bottom of the poll. And Macleod was certainly not seen as an obvious crofting spokesman.

But Angus Macleod possessed one asset which his predecessors had arguably lacked. He knew how to run a business. On becoming chair-man of the Lewis and Harris Crofters Union in 1977, largely because nobody else would agree to take the job, he began systematically to apply business methods to union organisation. He called regular meet-ings. He placed subscription-collection and record-keeping on a wholly new footing. By the early 1980s, helped by the resurgence of interest in crofting which accompanied the commencement of the Western Isles Integrated Development Programme, union membership in Lewis and Harris was well in excess of 1,000.

This success, paradoxically, served mainly to convince Angus Macleod that the Crofters Union movement, as then constituted, was inherently incapable of exercising worthwhile influence on behalf of crofters. His own union, Macleod realised increasingly, had neither the funds nor the personnel required to conduct effective lobbying in Inver-ness or Edinburgh where its leverage over organisations like the Crofters Commission and the Department of Agriculture was confined to the very limited pressure that could be exerted by means of the letters he himself sent occasionally to the officials dealing with crofting adminis-tration. What was really required, he concluded in the later 1970s, was to revive the Federation of Crofters Unions – and to equip the Federa-tion with a full-time organiser whose job it would be both to put the crofting case effectively to the government and to ensure that there was a substantial Crofters Union membership in every part of the Highlands and Islands.

This was the conclusion which Brian Wilson had reached, quite independently, at much the same time. Soon the *West Highland Free Press* was to be giving such support as it could to Angus Macleod's various endeavours.

Macleod began by tracking down the former Federation's funds and records – which, for lack of anyone else to assume responsibility for them, had passed at some point into the custody of the Assynt Crofters Union. With a view to formally reconstituting the Federation on as wide a basis as was practicable, he called a meeting in Stornoway to which he invited a number of Uist and Barra councillors who were then in town on local authority business. When that meeting failed in its objective, because nobody present would consent to accept the Federation chairmanship, a second gathering was held.

'This time,' Angus Macleod long afterwards recalled, 'I took the Federation chair myself. But that was only after I got the meeting to agree to making me secretary and treasurer as well. You see, I wanted office-bearers I knew would do some work!'[3]

Macleod now sought a meeting with Iain MacAskill, secretary to the Highlands and Islands Development Board. Would the HIDB help finance a revamped Federation, MacAskill was asked in November 1982. Such assistance was essential 'to prime the pump', as Angus Macleod put it in a paper he sent to MacAskill at that time. 'We would hope,' Macleod emphasised, 'after a while . . . to generate sufficient income to finance the project.'[4]

The Highland Board secretary did not rule out the possibility of some start-up cash being made available. There was a developmental case to be made for the creation of a more effective crofting organisation, he acknowledged. Angus Macleod, he suggested, should open more detailed discussions with Hugh MacLean, then in charge of the HIDB's Land Division.

MacLean – who would become Crofters Commission chairman in 1989 and who had himself grown up on a Tiree croft – proved to be not unsympathetic. In November 1983 he attended one of the meetings which the Federation had taken to organising in Inverness on the evening preceding the Commission's regular assessors conference and there announced that the Highland Board would be prepared to commission a feasibility study of the full-time organiser proposal. It was this study which, as mentioned previously, the present author conducted on behalf of the Federation of Crofters Unions in the early autumn of 1984.

In sharp contrast to what had gone immediately before, active support for the Crofters Union concept was clearly rising sharply by 1984 – with overall union membership possibly exceeding 2,000. Although the Lewis and Harris Crofters Union, now chaired by Gress crofter, Peter MacLeod, still accounted for more than half that total, the proportion belonging to other unions was beginning to grow steadily.

Efforts to resuscitate the Skye Crofters Union had begun with a public meeting at Sabhal Mor Ostaig in the spring of 1977. 'Everybody present was keen to get the union back into action again,' remarked the organisation's secretary, Willie Finlayson, a Scullamus crofter. By the early 1980s, with some help from South Skye regional councillor and Broadford minister, Revd Jack MacArthur, who gave the reconstituted union his strong personal backing, membership had reached 150.[5]

Similar developments were taking place in Uist and Barra. There the North Uist Crofters Union – thanks to the sometimes almost unaided efforts of office-bearers such as John MacDonald from Locheport and Alasdair MacDonald from Achmore – had been kept continuously in existence since its formation in the early 1960s. But its sister organisations to the south had long since fallen by the wayside. Now, with some encouragement from Comhairle nan Eilean's locally-based development officers, Neil MacPherson and Roddy MacDonald, the Benbecula, South Uist and Barra Crofters Unions were relaunched – the South Uist Crofters Union, for example, being formally reconstituted under the chairmanship of North Boisdale crofter, Roddy Steele, at a public meeting in Daliburgh in March 1980.

The Assynt Crofters Union, another survival from the 1960s, was holding its own – under the chairmanship of Culkein crofter, Frank Mirtle. The Lochaber Crofters Union had been reactivated by two local crofters – Ronnie Campbell from Bohuntin and Sandy Kennedy from Blarmafoldach. The Shetland Crofters Union, chaired by Laurence Graham from Veensgarth, was once again expanding. On the part of crofters in places with practically no tradition of Crofters Union activity – such as Wester Ross and the more easterly parts of Sutherland – there was no lack of interest in participating in any new organisation that might be launched.

Nor was there any shortage of opinion as to what a re-energised Federation of Crofters Unions ought to be about. It would not be sufficient, Roddy Steele insisted, 'for people to get on their hind legs once a year and talk grandly about the crofting way of life.' That would cut no ice with anyone – least of all with the genuinely active crofter.[6]

'A Crofters Union,' the South Uist crofter said, 'ought to consist of practising crofters talking about crofting and its development in a practical way.' There would have to be a preparedness and an ability – on the part of any new union, its office-bearers and its officials – to grapple

179

with the detail of public policy as it affected crofters; with agricultural price support mechanisms; with marketing prospects; with the levels of crofter housing grants and loans; with the perceived unfairness of many Crofters Commission decisions. Only by getting thoroughly to grips with issues of that sort, Roddy Steele and several others stressed, would a reformed Federation of Crofters Unions be taken seriously both by crofters and by the public agencies with which it would have to deal. But if they were indeed provided with effective representation, Roddy Steele enthused, crofters would gain immensely. 'We would begin to be consulted about policy,' he said. 'We would have to be consulted. We would have a chance of influencing decisions before they are taken instead of being left to protest uselessly about decisions after they have been made.'

It was the experience gained by Roddy Steele and his colleagues in the course of reconstituting their own unions in Uist which provided the 1984 report with its model for a wider reform of the Crofters Union structure. The chairmen and secretaries of the North Uist, Benbecula and South Uist Crofters Unions had been in the habit of meeting regularly in order to co-ordinate any necessary action on matters such as IDP administration. Thus they were able to combine the benefits of having a locally identifiable and locally responsible organisation with the equal advantage of centralised policy formulation. This approach, the 1984 report suggested, could usefully be applied more widely.

The new union, the report continued, should ideally have a substantial number of local branches – with places like Shetland, Skye, Lewis and Harris, for example, having several. To prevent undue fragmentation, however, each local branch would be represented, by two of its elected office-bearers, on an area committee of the kind already in existence in Uist. These committees would elect area presidents. Each area, of which there might be between eight and a dozen, would be proportionately represented on a policy-making central council – to whose members the union's full-time official, or director, would be responsible.

These proposals were put to a general meeting of the Federation of Crofters Unions in Inverness on 7 November 1984. They were carried overwhelmingly – with Roddy Steele boldly brushing aside any doubts. 'If we do not grasp this chance,' the South Uist crofter told his fellow delegates, 'we will not get another one. The present Crofters Unions will become weaker, not stronger; and for all that we will be able to achieve we might as well give up.'[7]

Existing Crofters Unions, it was agreed, would be asked, in effect, to disband themselves. And a steering group, with Angus Macleod as chairman and Ronnie Campbell from Lochaber as vice-chairman, would take charge of the many tasks which would clearly have to be accomplished if, as the 7 November meeting had insisted, a Scottish Crofters Union – the name soon given to the projected organisation – was to be put in place inside a year.

There remained one question to be answered. It was put to the Inverness meeting by a Gairloch crofter, Donald MacLeod. There had never been a Crofters Union in his area, he said. But there was interest in having such a union now. So would James Hunter be prepared to come and explain his report to Wester Ross crofters? Yes, came the response. He would.

So, on a wet and windy night at the beginning of December 1984, the possibility of establishing a Scottish Crofters Union was put – somewhat tentatively – to a gathering of Wester Ross crofters in Poolewe. Would such a union work? What would it do? What benefits would it bring? There were plenty of such queries. But there was real enthusiasm, too. Before the evening ended, the SCU had its first branch, to be known as Gairloch and District, its first branch chairman, Kenny Urquhart, its first branch secretary, Aimi Macdonald, its first branch treasurer, George Macleod, and, perhaps most significantly of all, its first subscriptions – of which there were no less than 38.

There were to be many more such occasions in the months ahead. In Lewis and Harris, where 14 branches of the SCU were eventually formed, Angus Macleod – assisted by Laxay crofter Iain MacIver, who was shortly to be appointed treasurer of the wider steering group, and by Comhairle nan Eilean development officer Roddy Murray – organised more than 30 public meetings and arranged for the distribution of several thousand copies of the leaflets which the steering group had printed in order to provide as much information as possible about the projected SCU.

The new union, readers of those leaflets were told, was being formed 'to promote and protect the interests of crofting communities throughout the Highlands and Islands'. The SCU would secure those objectives by:

ensuring that the case for crofting is put strongly and effectively to Government and to official agencies like the Crofters Commission and the Department of Agriculture; by taking up issues affecting union members locally and working hard to get these issues settled in favour of crofters; by uniting crofters everywhere in defence of the agricultural and other support services needed to maintain viable and worthwhile crofting communities; by providing members with a competitive and comprehensive insurance service; by publicising the crofting point of view; by campaigning for positive and constructive policies for the development and expansion of crofting.[8]

Such statements seemed to strike a chord with crofters. In Shetland, where the steering group was represented by Vidlin crofter John Stonehouse, SCU branches were formed on Unst, Yell and Fetlar – as well as on the Shetland mainland. Several union branches were launched in Skye – where Sleat crofter Farquhar MacLennan, aided increasingly by Waterloo crofter Angus MacHattie, took the lead on behalf of Angus Macleod's steering group. To the steering group's considerable surprise, there was intense interest in its activities in places which group members had begun by thinking to be outside their sphere of influence.

Caithness, East Sutherland and Argyll, the 1984 report had suggested, should be left, in effect, to the National Farmers Union of Scotland which was comparatively well organised in these localities. But now the steering group was being approached by people like Ardgay crofter Gregor MacKay, Tiree crofter Alex MacArthur and Islay crofter Deirdre MacLugash – all of them requesting the group to organise explanatory meetings in their areas. Similar appeals came from Morar crofter Ronnie Maclellan, Ardnamurchan crofter Allan MacPherson, Caithness councillor Bill Mowat and many, many more. The overall outcome was that the new union – instead of taking the several years envisaged by the 1984 report to put a branch network in place right across the Highlands and Islands – managed to get the greater part of the job accomplished in under 12 months.

'The level of support which the new Crofters Union has attracted has been nothing short of phenomenal,' one of Scotland's leading agricultural journalists commented in June 1985. 'In places where no support previously existed . . . hundreds of crofters have joined.'[9] There is no doubt that the demonstrable commitment of crofters to the SCU ideal was of great assistance to Angus Macleod and his steering group colleagues in their negotiations with the Highland Board – the more so as the group's financial ambitions grew with its success.

The total amount of cash at the disposal of the Federation of Crofters Unions in 1984 was £200. It is not surprising, therefore, that Angus Macleod's original suggestion that a reformed union should aspire to raise £10,000 annually had been met with no small amount of scepticism as to its feasibility. But now the steering group calculated both that the SCU would require an annual income of £72,000 and that the new organisation could be made self-financing, at that level of operation, within four years – providing that it received appropriate start-up aid from the HIDB.

These projections were to prove remarkably accurate. They took account of staff salaries, office overheads and the mounting of an annual conference. They also took account of the massive travel costs likely to be incurred by an organisation aspiring to serve some of the more inaccessible communities in the United Kingdom – an organisation which, for example, was eventually to have members on more than 30 separate islands.

If existing Crofters Union members, Angus Macleod explained to the Highland Board on the steering group's behalf, were to be levied at rates sufficient to provide the necessary cash, these rates would be so high as to make membership recruitment impossible. But if the SCU were made wholly dependent on such income as could be got from a lower subscription rate, the union would be unable to provide the high quality service which was clearly required if more members were to be attracted. Only the HIDB, it seemed to the steering group, could square that particular circle – by advancing the grants needed to enable the SCU, in its initial phase, to combine a good service with a reasonable rate of annual subscription.

These points, together with a draft budget and a draft union constitution, were put finally to the Highland Board in April 1985 – along with a formal appeal for financial assistance:

> The Crofters Union Steering Group believes that its achievements to date have been considerable. In the space of a few months, the entire Crofters Union movement has been reorganised and provided with effective central leadership and guidance. Dozens of new branches have been formed. Hundreds of new members have been recruited . . . . Now the Steering

Group looks to the HIDB to help finance the next stage in the most important development in crofting affairs for many years.[10]

In May it was announced that the Highland Board would provide the SCU with £38,500 in 1986, with £36,000 in 1987, with £18,000 in 1988 and with £14,500 in 1989. The HIDB, as Board vice-chairman Ronnie Cramond remarked a little ruefully on more than one subsequent occasion, had 'probably provided a stick for its own back'. But the Board, Cramond stressed, had been genuinely impressed by what was being created. And the steering group had thus got everything its chairman and its members had requested.

With Comhairle nan Eilean, Highland Regional Council, Shetland Islands Council, Strathclyde Regional Council and Ross and Cromarty District Council also making contributions of between £1,000 and £2,000 apiece, steering group members now had no doubts as to their ability to launch the new organisation on time. Although the group's chairman – recalling the difficulties he had experienced in getting even £200 together only the year before – was inclined to wonder privately just how the SCU of 1990 was to extract some 360 times the Federation's annual income from its crofting membership, he kept those anxieties carefully from the public and the press.

In October 1985 the Crofters Union Steering Group published the first issue of the SCU's quarterly newspaper, *The Crofter* – its title deriving from a Highland Land League publication of the same name. Crofting grants were being reduced and the agricultural advisory service in the Highlands and Islands was about to be curtailed, the paper commented. 'Crofting is under pressure. We need to respond. We need to be organised. And we need to be united.'[11]

With a paid-up membership already running into thousands, *The Crofter* commented in a front-page editorial, the SCU would be 'the largest crofting organisation any of us have ever seen'. The union, the new periodical stressed, was not promising to work miracles. 'But we now have an organisation run by crofters and in the interests of crofters. And we must make the most of that opportunity.'

An SCU director, the present author, had been appointed in the summer of 1985. Headquarters premises were shortly afterwards located in Broadford in Skye – the steering group being firmly of the opinion that the new union should be based not in Inverness, as were official agencies like the HIDB and the Crofters Commission, but in a crofting area. Towards the end of the year an SCU secretary and administrator, Fiona Mandeville, who came from a Skye crofting background, was recruited.

The SCU – which had been inaugurated formally in November 1985 at an Inverness meeting which also witnessed the winding up of the former Federation – became fully operational in January 1986. Angus Macleod, though installed as SCU honorary president, had insisted on taking no active office in the organisation he had done so much to bring into existence. The union's inaugural conference had accordingly elected as SCU president Frank Rennie – a crofter, a scientist and a naturalist from Galson in Lewis. Bohuntin crofter Ronnie Campbell was elected SCU vice-president. Joining the president and vice-president on the SCU Council were Iain MacIver, John Crichton and Alasdair Nicolson from Lewis, Alasdair MacEachen from Benbecula, Roddy Steele from South Uist, Alasdair MacInnes from Tiree, Angus MacHattie from Skye, Laurence Graham from Shetland, John Mowat from John O'Groats in Caithness, Alistair Fraser from Strath Halladale in North West Sutherland, John MacDonald from Rogart in East Sutherland and Angus MacRae from North Strome in Wester Ross.

Not the least striking thing about these men, in the context of a part of Scotland long characterised by its comparatively aged population, was their general youthfulness. The SCU Council's average age was under 40. A substantial number of its members – men like Frank Rennie, Iain MacIver, Alasdair Nicolson, Alasdair MacEachen, Alasdair MacInnes and Angus MacHattie – were in their twenties or their early thirties.

Of the SCU Council's 14 members, in fact, only two – John Crichton and Laurence Graham – had been actively involved in the first flourishing of the Crofters Union movement in the 1960s. And though a number of other veterans of that earlier period – notably John MacDonald from Locheport in North Uist, Hugh Matheson from Drumbeg in Assynt, Calum Nicolson from Portree in Skye and Margot MacGregor from Gartymore in Sutherland – helped greatly with the establishment of SCU branches in their various localities, the new organisation clearly belonged predominantly to people who were fresh to crofting politics.

One of the many unobtrusive ways in which the union contributed to the general advancement of crofting communities, the North Strome crofter and former Forestry Commission foreman, Angus MacRae, was to remark in 1991, towards the end of his own three-year stint as SCU president, was the extent to which the union had become 'the medium through which so much crofting talent has been given the opportunity to express itself'.[12]

It was most certainly the case that the SCU did not suffer from any lack of organisational ability on the part of crofters. Alasdair MacEachen, a Benbecula crofter and local government officer, was to get his union branch membership up to 100 per cent of what was possible. John MacDonald, a Rogart postman, crofter and danceband leader was to make East Sutherland one of the SCU's liveliest areas – organising ceilidhs and excursions as well as the more regular round of evening meetings. Donald Linton, a Glencruitten crofter and lorry driver, was to recruit 100 SCU members in a part of Argyll where the union's founders had not intended even to have a branch. Scores of others were to make a very similar contribution.

An SCU council meeting, of which there were four or five a year, might involve a Shetland or a Western Isles representative in a round trip that could not be accomplished in under three days. Union office-bearers from islands like Islay, Tiree, Barra, Raasay, Unst and Yell had to be away from home overnight if they wished even to attend a meeting of their area committee.

But the necessary journeys were almost always made; the necessary space for union business somehow found in lives that were already crowded. And that was clearly just as well. The SCU was never going to be so wealthy as to be able to dispense with voluntary effort. Had it not been for the quite astonishing willingness, on the part of so many men and women, to undertake the mundane and routine jobs which had to be done to set the union on its feet, the SCU would very soon have foundered.

Organisation, especially financial organisation, had been Angus Macleod's great worry. There had been some suggestion, he noted in February 1985, that a fixed proportion of each SCU subscription should be retained at branch and area level. But no such approach, Macleod insisted, should be tolerated for a moment. Were union funds made the property of local branches, he forecast, branches in places where crofters were few would soon go short of cash – while money would be accumulated uselessly in other localities, such as Lewis, where crofters were more numerous.

'The strong should help the weak,' Angus Macleod commented. But for this to be accomplished successfully, control of union finances would have to rest, from the first, with SCU headquarters. 'The money collected for the new union should go to the top and central body,' Macleod ruled. The SCU headquarters, he continued, should also be primarily responsible for the collection of subscriptions.[13]

Addressing the question as to why Crofters Unions had so often failed to endure before, Macleod concluded, bluntly but perceptively, that they had gone under, in practically every case, 'because there was not a proper system of collecting union dues'. Unless the SCU quickly devised and implemented some such system, Angus Macleod warned sternly, it would – for all the HIDB's generosity – very rapidly go the same way.

It was this duty that was to devolve on SCU administrator Fiona Mandeville. With the assistance of her husband Geoff, a computer expert who gave freely of his time and skill to produce the necessary software, she devised a computerised membership register which enabled the union to maintain regular contact with each and every SCU member. Branch membership lists – giving details of all outstanding subscriptions – were sent regularly to branch chairmen, secretaries and treasurers. In order to lessen the burden of subscription-collection on those local office-bearers, subscription reminders were posted to members individually.

A high priority was given to a campaign to persuade SCU members to pay subscriptions directly from their bank accounts by means of variable direct debit. Eventually more than half the total membership had signed the necessary mandates. So successful was Fiona Mandeville's organisational streamlining that, in December 1990, she was able to report to the SCU council that more than two-thirds of all the union's members had paid their 1991 subscription in advance of the commencement of the calendar year to which that subscription actually applied.

Subscription rates had caused much anguish from the start. The locally-based unions which the SCU replaced had charged their members no more than one or two pounds annually. The Crofters

Union Steering Group, in the course of 1985, had requested £4. This was raised to £8 in 1986, to £11 in 1988 and to £16 in 1990 – the year in which the SCU was required to manage entirely without Highland Board subventions. These, admittedly, were not large sums. In relation to previous Crofters Union practice, however, they seemed substantial enough. Each increase was widely expected to result in a loss of membership.

But this did not happen. Indeed the SCU expanded more rapidly than had been foreseen by even the most determinedly optimistic members of the 1985 steering group. The group's publicly stated aim was to recruit an SCU membership, by 1990, of 4,000 – that being the number which had to be inserted into the financial projections made available to the HIDB if those projections were to demonstrate that the SCU would, one day, be self-financing. But steering group members were well aware that the 1984 survey had found a Crofters Union membership of not much more than 2,000 – a membership, moreover, which was in the habit of paying subscriptions at rates far below those which the SCU would have to levy. It was understandably difficult to be sure that existing union members would be retained, let alone to be confident that another 2,000 would be found.

But found they were. SCU membership passed 3,000 in 1986 and 4,000 in 1987. And though subsequent expansion was rather less spectacular, there continued to be net annual increases – with paid-up membership, in December 1990, standing at 4,419.

A major factor in this impressive growth was the SCU's success in establishing itself outside those localities traditionally identified with Crofters Union activity. Of the 2,000 or so people belonging to Crofters Unions in 1984, after all, almost two-thirds lived in either Lewis or Harris. But Lewis and Harris, by 1990, accounted for under one-third of SCU members. On the Highland mainland, where Federation of Crofters Union affiliates had under 100 members in 1984, the SCU had well over 1,000 members six years later.

The speed at which the SCU attained its membership targets helped the organisation to attain its financial targets also. Membership subscriptions were bringing in some £70,000 annually by 1990. An associate membership – consisting of local authorities, commercial concerns and a loyal, if very widely scattered, band of sympathetic individuals – was contributing a further £4,000. Bank interest – generated mainly as a result of Fiona Mandeville's highly successful efforts to ensure that SCU members paid subscriptions promptly – was worth another £6,000. Considerable advertising revenues were being generated by *The Crofter*. Sales of items as varied as mugs, ties and sweatshirts were resulting in worthwhile earnings. A number of island, regional and district councils were continuing to provide an annual total of some £4,000 in grant aid. The SCU, in short, was in a reasonably secure position.

Not the least significant element in accounting for the strength of the various farming unions in the United Kingdom has been the longstanding relationship between those unions and the NFU Mutual Insurance Society. The society was formed in 1910 when seven Warwickshire farmers, who had met in a Stratford-upon-Avon teashop in order to discuss possible means of persuading their fellow agriculturalists to join a local farming union, came up with the notion of making union membership the one means of access to the services of the agriculturally-orientated insurance provider which they then proceeded to establish.

By 1919 the organisation thus created had become official insurer to the National Farmers Union of England and Wales. By 1930 it had established a similar connection with both the National Farmers Union of Scotland and the Ulster Farmers Union. Today, though the NFU Mutual has gone on to become one of the United Kingdom's major insurers, it still fulfils the principal objective of its founders – by making farming unions considerably more effective than might otherwise be the case.

Not only is membership of a farming union the essential passport to the Mutual's ever-expanding range of financial and insurance services, the Mutual's local agents are also union branch secretaries. As such they receive a salary of a sort from the union which they serve. But that salary is entirely nominal – the secretary-cum-agent depending for the bulk of his livelihood on the commissions earned from his insurance dealings on behalf of the Mutual. The result is that, at minimal cost to organisations such as the NFUS, a nationwide network of suitably equipped and generally well staffed offices is available to cater to the requirements of those farmers who belong to one of the major farming organisations.

The advantages of the SCU making its own connection with the Mutual – or, failing that, with some comparable concern – were

189

recognised in the 1984 report on Angus Macleod's reform proposals. Negotiations with the Mutual were accordingly begun by the Crofters Union Steering Group in 1985 and concluded by the SCU itself in the early part of 1986 – the various discussions between the two parties being helped by the explicit recognition, on the part of Vince Seaman, assistant general manager at the Mutual's Stratford headquarters and the man in charge of hammering out an appropriate deal, that the SCU was then at a stage of its development not altogether dissimilar to that of the Warwickshire farming group at the time of its membership's decision to make the original link between agricultural unions and insurance.

SCU members, it was agreed in 1986, would have access to NFU Mutual services on exactly the same basis as NFUS members. In much of the Highlands and Islands these services would be supplied to SCU members by existing NFUS branch secretaries. But in the Western Isles the SCU would appoint two secretaries of its own – one to service the union's Lewis and Harris area committee, the other to do the same job in respect of the Uist and Barra area committee. Both those individuals would handle NFU Mutual insurance business in their own right. The Mutual would guarantee them a basic income over the two years it was thought likely they would require to establish themselves.

The two area secretaryships were advertised in the summer of 1986. The Uist post was occupied first by Martin Matheson and later by Mary Johnson – both young local people. The Lewis and Harris job went to Angus Graham – a councillor, a tweed weaver, a former trades union activist and a staunch Labour Party man who turned out to have the knack of being able to combine a commitment to community service with a quite outstanding business acumen.

Both Vince Seaman and the Mutual's Glasgow-based Scottish manager, Ken Smith, had expected their newly established Western Isles agencies to be part-time positions. That was the pattern set, over many years, by NFUS branch secretaries in comparable localities – most notably Shetland. By the end of the 1980s, however, Angus Graham was not only winning NFU Mutual salesmanship awards; he had also built up an insurance operation so substantial as to be employing himself, his wife Isobel and an office junior.

Mary Johnson in Uist was also working full-time on SCU and NFU Mutual business. And since the union – in addition to its two

190

headquarters staff in Broadford – was also employing an Ollaberry crofter, Drew Ratter, as its part-time area secretary in Shetland and a Waternish man, David Gillies, as its part-time area secretary in Skye, there was a sense in which the organisation could be said to have created several jobs in places where even such small-scale expansions in employment are important.

But much more significant, from the SCU perspective at any rate, was the extent to which the NFU Mutual connection made it possible for the organisational structure which had so admirably served the United Kingdom's farming industry to be adapted to a crofting purpose. It was not simply that its painstakingly negotiated agreement with the Mutual enabled the SCU to associate union membership with a demonstrable financial benefit – though that was certainly helpful. Nor was it merely that the Mutual assisted the SCU to produce attractive brochures and other literature designed to assist the union in its recruitment efforts – though that was also valuable. It was rather that the SCU, because of its link with the NFU Mutual Insurance Society, was able to maintain two regularly staffed offices in that most important crofting locality, the Western Isles. No such permanent presence in either Stornoway or Balivanich could ever have been financed out of membership subscriptions.

It was the SCU's good fortune to be launched in the centenary year of the Crofters Act of 1886 – a circumstance which resulted in the union and its aspirations being widely featured in the various television documentaries, radio broadcasts, newspaper articles and historical exhibitions generated by this notable crofting anniversary. The SCU was not slow to draw its own conclusions from the events which were thus celebrated:

> The Crofters Act was not granted willingly. It was the product of determined action on the part of crofters and their families. We owe the Crofters Act, and the survival of our crofting communities, to our ancestors who refused to be intimidated by the landlords who removed them from their homes and the Governments which deployed the police and the military against them. Above all, we owe the Crofters Act to the leaders and members of the Highland Land League who provided our people with the means to triumph over their oppressors.[14]

The SCU, this union statement continued, was conscious of that inheritance and of its duty, 'as the largest crofting organisation since the Highland Land League', to build constructively on the foundations laid so enduringly by its predecessor. 'Like the Highland Land League,' the SCU council commented in June 1986, 'the Scottish Crofters Union will endeavour to uphold the right of crofting families and communities to shape their own future and to live their own lives in ways that seem best to them. And like the Highland Land League, the Scottish Crofters Union will insist that the determination of crofting policy is primarily a matter for crofters.'

But how exactly was this to be accomplished? The SCU was extending new insurance services to crofters. With the help of several Highlands and Islands solicitors – most notably Derek Flyn, an Inverness-based specialist in crofting law – the union was also providing its members with a legal advisory service which, over the next few years, would secure a number of very satisfactory settlements on issues ranging from the level of croft rents to the compensation due to crofters whose land was required for non-agricultural purposes. But for all that those were welcome developments, they did not, of themselves, alter the context in which policy for crofting was determined by government and its various agencies. If the SCU was to prove itself capable of making a wider impact, it was clear, the union would have to get seriously to grips, as the 1984 report had stressed, with a wide range of crofting policy issues.

A start was made with the Department of Agriculture's crofter housing scheme – the mechanism which permits public money to be advanced to crofters, as has been the common practice for much of the twentieth century, in the form of both grants and low-interest loans for housing purposes. This type of assistance had long been regarded with suspicion, even hostility, by many folk who were not crofters. Although the SCU was convinced that such aid could, in fact, be shown to be highly cost-effective, it seemed essential to the union council that this be proved once and for all – such proof being all the more urgently required in view of the extent to which so many public spending programmes were being curtailed by Margaret Thatcher's Conservative government.

In the summer of 1986, therefore, the SCU council established a housing sub-committee, under the chairmanship of Roddy Steele, 'to make a detailed investigation', as was stated in the relevant union press release, 'of the Crofter Building Grants and Loan Scheme operated by the Department of Agriculture and Fisheries for Scotland'. To help them with this task the sub-committee's members engaged the services of Aberdeen University housing expert Mark Shucksmith.[15]

One immediate outcome was the appearance, in January 1987, of *Crofter Housing: The Way Forward*, the publication in which the SCU summarised its housing policy recommendations. This booklet, as Roddy Steele commented in his introduction to it, made 'a powerful and well-supported case' for the continuation of public investment in crofter housing. It made an equally powerful case, in Roddy Steele's opinion, 'for the existence of the Scottish Crofters Union' – in that there appeared to be no other organisation with the capacity, or the willingness, to undertake the defence of the crofter housing scheme.[16]

That scheme, the SCU report made clear, certainly required reform. But such reform, the union argued, should be designed to make it more, not less, widely available. Certainly there was no financial justification, in the SCU's considered judgement, for reducing Department of Agriculture expenditure on crofter housing.

The public expenditure implications of building a new local authority house on a Hebridean island, the SCU's 1987 report demonstrated, were more than three times greater than the cost to the taxpayer of providing a crofter on the same island with his maximum entitlement to crofter housing grant and loan assistance. Nor was the Treasury any more disadvantaged by the financing of those grants and loans than it was by successive governments having extended generous tax relief, on an almost universal basis, to those people who chose to purchase or construct their homes with the help of money borrowed from banks or building societies.

'The crofter housing scheme has helped retain population in the countryside,' the SCU was to observe in a 1989 summary of its position on the housing issue. 'It has enabled many families, who would not otherwise have been able to do so, to provide themselves with their own homes. It has secured these benefits at minimal costs in terms of public expenditure.'[17]

These were key points to make to a cost-cutting Conservative administration which was publicly committed both to the maintenance of well-peopled rural communities and to the widest possible expansion of home-ownership. They constitute a fairly typical example of the means by which the SCU sought to break with the longstanding tradition of basing the political case for the continuation of crofting on largely sentimental considerations.

'There is no point in simply shouting about crofting,' said SCU president Frank Rennie at one of the union's annual conferences, 'no point in simply getting misty-eyed about crofting. We have to make a solidly based and highly detailed case in support of our objectives. And we have to make the case for crofting in terms that make sense to the present Government.' [18]

This was all the more essential in view of the fact that the 'present Government' – with whose ministers Frank Rennie and the SCU were having ever more regular dealings as their claim to speak for crofters was increasingly accepted by the Scottish Office – consisted of hard-headed Conservative politicians who were most unlikely to be impressed by any amount of woolly talk as to the alleged spiritual or moral merits of the crofting existence. So the SCU set about the elaboration of a more pragmatic set of arguments as to why government should take a constructive interest in crofting affairs.

'On a purely statistical basis,' the SCU director observed in *The Crofter* in the course of the summer following the 1987 general election, which confirmed the Conservative Party's national grip on power but which also saw the Western Isles fall – after some years of SNP predominance – to Labour's Calum Macdonald,

> the Western Isles ought to be a Conservative stronghold. Because of crofting's dominant role in the local economy, the proportion of the electorate involved in agriculture is higher in the Western Isles than in any other parliamentary constituency in Britain. Also because of the dominance of crofting, rates of home-ownership and self-employment are far higher in the Western Isles than in most of Scotland. These, of course, are characteristics which are generally thought to make a community more likely to vote Conservative. In the June general election, however, the home-owning, self-employed agriculturalists of the Western Isles returned a Labour MP and the Tory candidate, as is usual in the islands, came bottom of the poll. Why? [19]

One part of the explanation, the SCU director suggested, was to be found in the extent to which Margaret Thatcher's Conservative Party continued to be almost wholly identified, in a north of Scotland context, with absentee landlords and City of London forestry companies rather than with the resident population of the area.

> Thatcherism is supposedly on the side of small business and the independent-minded citizen. And it would be much more in tune with those much proclaimed objectives if the Conservative government, in its Highland aspect, was associated rather less with the maintenance of sporting estates or tax-avoiding afforestation and was identified rather more with promoting the interests of the crofter-fisherman, the crofter-weaver and all these other small-scale island entrepreneurs who, in June, clearly felt they had more in common with Socialism than with Conservatism.

This 1987 article – somewhat to the surprise of its author who had also made the point in the course of it that previous Conservative governments had not been averse to embarking on radical policy initiatives in crofting areas – was promptly seized upon by Lord Sanderson of Bowden, the Borders businessman whom Margaret Thatcher had just then appointed Scottish Office Minister of State with responsibility for both the Highlands and Islands and agriculture.

'For what it is worth,' Lord Sanderson told the SCU's 1989 annual conference in Benbecula, 'the only far-reaching land reforms in the Highlands this century, the land acquisition and land settlement measures of the 1920s, were the work of a Conservative or Conservative-dominated administration. I stress that the principles of my party and government, which believe in home-ownership and self-employment, are also, I believe, the aspirations and wishes of the crofting community.'[20]

Although the expression of such sentiments did not, of itself, imply that the Minister of State was about to grant the Scottish Crofters Union all its wishes, there was no doubt as to the Scottish Office's growing willingness to enter into genuine dialogue with the SCU delegations which first Frank Rennie and then Angus MacRae were now leading regularly to Edinburgh.

A particular union grievance concerned the regulations governing the payment of suckler cow premium – one of the more important elements in the support available to beef cattle breeders. Worth some £33 a head by 1988, and due to rise sharply as a result of European Community beef régime reforms which were then imminent, the premium was confined to those beef producers who earned more than half their total income from agriculture. This, of course, meant that it could not be claimed by the many part-time crofters whose earnings were derived principally from their non-agricultural occupations.

Some two dozen SCU members who volunteered the necessary personal details to the union were shown to be losing about £6,000 annually between them as a result of what the SCU had labelled – perhaps a little hyperbolically – this 'blatant unfairness and discrimination'. The suckler cow premium issue, as the Minister recalled in the course of his Benbecula speech, consequently loomed large at SCU meetings with Lord Sanderson.

'I was left in no doubt,' Sanderson told delegates to the union's third annual conference, 'about the importance of the changes in the beef régime which were being contemplated as far as the crofting community was concerned. You were asking us to press to secure the removal of the income test arrangements which have denied so many crofters the support of beef production available under the suckler cow premium. We delivered this. I understand that as many as 1,200 crofters and farmers stand to benefit from the change.'[21]

Not every SCU campaign ended so positively, of course. The union failed, as did many other much more powerful organisations, to dissuade the Conservative government from replacing the local government rating system with the community charge or poll tax – a reform which inflicted an especially damaging financial blow on crofting communities because those communities, as a result of concessions made to them by previous Tory administrations, enjoyed partial exemption from domestic rates. Nor did the SCU obtain as much as its president and council would have liked from the annual negotiations concerning the level of agricultural support payments – or hill livestock compensatory allowances – made to sheep and cattle producers in more disadvantaged localities such as the Highlands and Islands.

But the fact that crofting representatives had finally gained access to these discussions – which had gone on without any crofting participation since the introduction of the original hill sheep and hill cattle subsidies in the 1940s – was itself no small achievement. The SCU, at any rate, was convinced that such gains as were made in respect of compensatory allowance disbursements in the north of Scotland in the later 1980s could legitimately be attributed, in part at least, to the increasingly well-researched submissions with which union negotiators were able to substantiate the claims they made on behalf of their members.

Union pressure for something to be done to give mainland crofters access to benefits of the kind accruing to their island counterparts as a result of the IDP and the ADP helped produce both the Highland Board's North West Development Programme and the subsequent, much more widely applied, Rural Enterprise Programme. SCU insistence that crofters be given ownership rights to trees planted on their common grazings – such trees being traditionally considered the property of crofting landlords rather than crofting tenants – was to result, in 1991, in the necessary reforming legislation being placed before parliament by Western Isles MP Calum Macdonald.

The union, a little to its council's surprise, found itself getting on fairly well with those Scottish Landowners Federation representatives – foremost among them Lord Strathnaver of the Sutherland Estate – with whom the SCU, under Angus MacRae's leadership, thrashed out a mutually-agreed formula on how to enable crofting townships to establish woodlands on their common pastures. But between the SCU and the Crofters Commission there was rather more in the way of friction – the union being convinced, as a result of its membership's experience of Commission decisions on matters such as tenancy assignations, that the Commission was not always acting in ways calculated to serve the wider crofting interest.

Widespread dissatisfaction with the Commission's handling of assignations, decroftings, apportionments, crofter absenteeism and related issues was nothing new, of course. Such discontent had contributed substantially, as mentioned in an earlier chapter, to that crofting disenchantment with the Crofters Commission which became so spectacularly evident in the years around 1960. There was a very real sense in which the Commission, as also mentioned previously, simply could not go about its regular business without displeasing one side or the other in the many disputed cases it was statutorily obliged to resolve.

What changed with the emergence of the SCU was that it was now much easier than previously for those crofters – and there were no small number of them – who believed themselves to have been treated unjustly by the Crofters Commission to find a means of publicly pursuing their various grievances. The consequent scope for disagreement

197

between the union and the Commission was made all the greater by the fact that it took both the SCU and the Crofters Commission some considerable time to establish satisfactory working relationships with one another.

The SCU – especially in its bolder and brasher early days – did not hesitate to accuse the Commission, for example, of 'breaching its statutory responsibilities' by permitting substantial numbers of crofts to be removed from crofting tenure by their owners. Such comments caused offence, and that offence was all the greater as a result of the Crofters Commission – possibly because of the secretive and legalistic tendencies engendered by its quasi-judicial responsibilities – clearly finding it much harder than other comparable organisations to establish with the union those informal links which commonly enable pressure groups of every kind to deal satisfactorily with officialdom.

With the Highland Board and the Department of Agriculture, the SCU had good relations from the outset. With the Crofters Commission it did not. When, in 1987, the SCU council published a highly critical discussion paper on Commission policy – the paper being principally the work of the union's director and its North West Sutherland area president, Bill Ritchie – the ensuing furore was such as to persuade the SCU that it had better try to behave more circumspectly in the future.

That the Crofters Commission chairmen with whom it dealt in the course of its first five years – the former agricultural adviser, Archie MacLeod, and the former HIDB man, Hugh MacLean – were personally determined to work as closely as possible with the SCU was something that the union never questioned. It was undoubtedly the case that, after the 1987 fracas, relations improved steadily. Regular meetings of union and Commission representatives were initiated. Other contacts grew. Union concerns began to be reflected in Commission policy documents. And though the SCU's first director always felt that this was the area of union business he had coped with least successfully, possibly because his journalistic inclinations always made it hard for him to resist publicising a good-going quarrel, the SCU, by the early 1990s, had clearly been accepted, even by the Crofters Commission, as having a vital role in crofting matters.

It had always been Angus Macleod's ambition to make the SCU 'part of the furniture', as he habitually put it. Media comment as to the sheer achievement of having got the union up and running was no doubt very gratifying, he would explain. But nobody ever made similar remarks with regard to the existence of the National Farmers Union or the National Union of Mineworkers. They were simply taken for granted. It was devoutly to be wished, in Angus Macleod's opinion, that the SCU would one day be taken for granted also; that it would eventually seem wholly natural for crofters, just like other groups of people, to have automatic access to a representative organisation with the clout needed to protect and to advance their interests.

Crofting history, as Angus Macleod was only too well aware, offered no great comfort in this regard. The better part of a century separated the Highland Land League from the Federation of Crofters Unions. But both, for all that they had commenced by attracting widespread backing and for all that they had earned the right to that backing by their able advocacy of the crofting viewpoint, had eventually failed more or less completely; their leaders at odds with their followers; their funds exhausted; their support dissipated; their influence, for all practical purposes, entirely at an end. So why should the Scottish Crofters Union do any better?

That the SCU was by no means incapable of antagonising some sections of crofting opinion was demonstrated by arguments as to how the union should react to government suggestions, made in the early part of 1990, that the ownership of the Department of Agriculture's crofting estates in Skye and Raasay should be transferred to a locally elected crofting trust.

The 100,000 or so acres in question comprised lands acquired by the state for settlement purposes in the years between 1900 and 1930. Fearing that the Skye and Raasay estates might be sold to private landlords by Conservative ministers committed to the wholesale disposal of publicly-owned assets, the SCU urged acceptance of the government's offer. But the community ownership approach, though endorsed unanimously by the union council and by the SCU's Skye and Lochalsh Area Committee, was regarded less enthusiastically by many of the directly affected crofters. Community ownership was certainly preferred to any sale of Department of Agriculture land to private interests. But the SCU, a number of crofters clearly felt, should have fought rather harder for a retention of the traditional relationship between the Department and its crofting tenants.

Among younger crofters, in particular, there was support for the concept of crofting communities taking direct responsibility for the management of their own affairs. And for all the criticism levelled at the union in the course of the community ownership debate, few Skye or Raasay crofters showed any inclination to give up their membership of it. That was encouraging.

As the SCU's new director, George Campbell, the young man who had previously been in charge of the HIDB's North West Development Programme, began the travels which, in the course of 1990 and 1991, would take him to every corner of the Highlands and Islands, there were other grounds for hoping that the Scottish Crofters Union, unlike the Land League and the Federation, would not quickly fade away.

Union membership, five years after the SCU's launch, was still tending to increase. Union funds, though not unlimited, were proving sufficient to keep the organisation well afloat. Also, the wider national and international context, in which the SCU was increasingly required to operate, was becoming steadily more favourable from a crofting point of view.

Chapter Eight

THE CASE FOR CROFTING

In the summer of 1898, some months after the establishment of the Congested Districts Board which had been formed by Lord Salisbury's Conservative government in order to promote the economic development of the crofting areas, the Landward Committee of Stornoway Parish Council, in the person of its clerk, J. F. MacFarlane, suggested to the new organisation that it might appropriately embark on a land settlement scheme in the vicinity of Stornoway where, in the committee's opinion, the farms of Aignish, Holm, Gress and Sandwick could readily be divided into smallholdings.

The farms in question, the Stornoway committee considered, were 'admirably suited' for settlement by 'the fisher-cottar class'. The committee, explained MacFarlane, had consequently instructed him to ask Congested Districts Board members to 'exercise their legal powers on behalf of this hard-working and deserving class by subdividing land among them for cultivation and grazing'. Such subdivision, Mac-Farlane's letter further urged, should be carried out in such a way as to produce as many crofts as possible of between two and four acres – this being sufficient to permit a man to combine the management of a piece of ground with his work in the fishing industry.[1]

To practical people in Lewis this doubtless seemed an eminently sensible request. But that was not how it was viewed by staff at the Congested District Board's Edinburgh headquarters. 'The Board,' ran the curtly dismissive note soon sent to Stornoway, 'consider that . . . no new agricultural holding should be created by outside assistance unless it is of such extent that its occupier could depend entirely, or at any rate mainly, upon it for a fairly comfortable living. The size of croft suggested by the Landward Committee is, without doubt, considerably below this standard.'

The sheer force of crofting opinion was eventually to oblige the

Congested Districts Board to modify its position. And the farms mentioned by the Stornoway committee – like most similar farms elsewhere in the Western Isles – have now been in crofting occupancy for very many years. But the Board's initial stance was never abandoned completely. In its evident antipathy to the smaller croft, Congested Districts Board policy was wholly in line with the longstanding conviction – on the part of successive generations of civil servants, officials and administrators of every kind – that the crofting system was essentially inefficient, outmoded and generally incapable of adaptation to the requirements of a modern economy.

Such, for example, was the considered judgment of the Napier Commission which was charged with the task of examining the causes of the crofting unrest which so convulsed the Highlands and Islands in the 1880s. That commission was not unsympathetic to crofters. Its members were well aware of the suffering caused by the clearances. They recognised the injustice implicit in the previously unfettered right of landlords to eject crofters from their holdings. They also advocated the introduction of security of tenure. But they wished to grant such security only to the occupants of bigger crofts – the tiny number with access to sufficient land to enable them to be full-time agriculturalists.

Lord Napier and his colleagues, then, believed the cause of crofting problems to lie in the crofting system's failure to evolve into the highly specialised farming structure which, even then, was characteristic of the greater part of Britain's countryside. In seeking to deny security of tenure to those crofters whose agricultural efforts were necessarily conducted in association with an almost endless array of other activities, the Napier Commission was simply endeavouring to facilitate the amalgamations which alone could transform part-time crofts into full-time farms.

Napier's intentions, of course, were frustrated by the Highland Land League's successful insistence that parliament should extend the security of tenure provisions of the first Crofters Act of 1886 to every crofter – however diminutive his holding. But to those numerous observers and commentators who shared the Napier Commission's conviction that any society founded on a smallholding system was intrinsically incapable of advancement, this was an historic error.

'It is my belief,' wrote the economist Farquhar Gillanders in 1968, 'that the 1886 legislation heralded the doom of crofting as a way of life, insulating it almost completely from normal economic trends and legally ensuring that crofting land could not now be developed into viable economic units.'[2] For much of the last 100 years, for all the crofting community's stubborn refusal to surrender what had been so bravely won in 1886, most received wisdom was undoubtedly on Gillanders' side.

Those junior civil servants in charge of the actual implementation of land settlement in the period between 1900 and 1930 were frequently committed strongly to what they were about. But more senior administrators openly derided the proliferation of publicly funded smallholdings – clearly regarding the entire settlement exercise in the north of Scotland as 'a terrible mistake'.[3] Throughout the 1940s, 1950s and 1960s the Department of Agriculture, when not actively engaged in attempts to undermine crofting tenure, was religiously rejecting all the many efforts made to have crofting treated as something other than a hopelessly ineffectual variant of farming.

'The error that can be traced throughout the whole course of crofting legislation,' commented Donald J. MacCuish, for many years both solicitor and secretary to the Crofters Commission and the leading modern authority on crofting law, 'is the treatment of the crofting problem as an agricultural problem capable of an agricultural solution.'[4] Indeed it was the Department of Agriculture's determination to persist with just such a policy that was to result in the Crofters Commission, in the course of MacCuish's service with it, becoming so disastrously embroiled in yet another effort to propel crofting down the road which Napier had mapped out so long before.

Had the Scottish Office, when piloting through parliament the measure which eventually became the Crofters Act of 1955, been at all sympathetic to the argument – advanced cogently and persuasively by MPs like Malcolm MacMillan and Jo Grimond – that the key to expanding the crofting economy was to provide the means of developing those 'ancillary industries' to which the Taylor enquiry had earlier devoted so much of its time, then the Crofters Commission would have had a much better chance of assisting the part-time crofter to expand his income. At the Department of Agriculture's insistence, however, the financial aids and incentives at the Commission's disposal were limited strictly to the agricultural arena. And when Sir Robert Urquhart and his colleagues decided, in 1960, to advocate the wholesale amalgamation of crofts, they were arguably doing no more than bowing to the logic of the situation in which they had been placed by the Act of 1955 and the subsequent Crofting Counties Agricultural Grants Scheme – both of which were all too clearly stamped with the Department's imprint.

The amalgamation proposals of 1960 were to be defeated by the Federation of Crofters Unions in much the same way as the Napier Commission's very similar proposals had been defeated by the Highland Land League. The Federation was duly castigated by Farquhar Gillanders – whose hostility to crofting attitudes was of that peculiarly intense sort which is only to be found among those who, like the Applecross-born Gillanders, feel that they have somehow shaken off the shackles fastened on them by their own crofting background – in terms depressingly akin to those employed by the many landlord critics of the original Crofters Act.

What was being urged by the Crofters Commission in 1960, wrote Gillanders, a Glasgow University lecturer, was 'the long-awaited ruthless reappraisal that alone would bring economic progress to crofting'. But emotion and irrationality, Gillanders thought, had once more triumphed over common sense. Crofters clearly preferred 'stagnation, apathy and economic ignorance' to the 'rationalisation of crofting agriculture'. In consequence, crofting townships were destined to remain forever 'sterilised in an outdated pattern of minute units'.[5]

This was powerful stuff. It exerted an undeniable appeal at a time when United Kingdom agricultural policy, as already noted, was concerned almost exclusively with maximising food production; with making farms ever larger; with mechanisation; with intensification; with the application to farming of every possible variant of science and technology. Because Gillanders, rather than Charles MacLeod, Malcolm MacMillan and other defenders of the part-time holding, seemed so much in tune with the spirit of the age, the case which the economist advanced indubitably helped to foster an attitude of mind to the effect that to be a crofter was, almost by definition, to be incapable of full and proper participation in the wider society of which crofting communities were a part.

From the perspective of the 1990s, however, it is the Gillanders approach to economic development – not that of his Crofters Union opponents – which looks increasingly at odds with the temper of the times. Today the European Community, including Britain, is entirely self-sufficient, indeed more than self-sufficient, in most temperate food-stuffs. The current agricultural policy priority is consequently to curtail, not expand, production. A closely related objective – as the farming support régimes which were first put in place in the 1940s begin to be withdrawn and as agricultural profitability declines in consequence – is to find new sources of farm income; to diversify the rural economy; to make country communities less dependent on agriculture.

'What is needed,' commented the European Commission in one of the many documents signalling this fundamental change in thinking, 'is not only action regarding agriculture itself but also a policy for creating lasting, economically justified jobs, outside the farming sector.'[6]

Such comments, of course, were the common coin of the much derided Crofters Union activists and pro-crofter politicians of the 1960s; of the many men and women who rallied behind the successful campaign to dish the Crofters Commission's plans for the elimination of the part-time holding; of that small but prescient band of crofting enthusiasts – prominent among whom was John Rollo of the Highland Fund – who insisted that, contrary to the overwhelming weight of orthodox opinion, it made sense to plan for a future in which industrial employment and part-time farming would widely complement each other.

Rollo would have been gratified by the European Commission's having discovered, a little to its surprise, that 'more than 60 per cent of new jobs created in Italy in the 1970s were located in rural areas'. He would have welcomed the fact that 'industrial employment also increased in rural areas of France'. And he would perhaps have felt his own ideas vindicated most of all by the finding that in Germany, one of the world's most successful economies, 'a fairly balanced geographical distribution of economic activity undoubtedly helped to stimulate strong growth of part-time farming supplemented by off-farm work' – to the extent that 'almost half the farms in Germany are run on a part-time basis'.[7]

But among all the many rural policy reversals which have occurred in recent times there is one, it is fairly safe to say, which would have given particular pleasure to John Rollo, to Frank Fraser Darling, to Malcolm MacMillan and to all those other maverick characters – many of them mentioned in earlier chapters – who battled so tenaciously with

205

an unyielding Department of Agriculture on behalf of the proposition that crofters ought to be the beneficiaries of measures designed to add to their non-agricultural earnings. In 1991 the Department of Agriculture itself launched, in selected crofting areas, its Rural Enterprise Programme – the latest in the series of very welcome development initiatives which the European Community has helped finance in the Highlands and Islands. REP cash aid, said Scottish Office ministers and civil servants, with all the insouciance of men who have just come up with something wholly novel, would be earmarked largely for non-farming projects of a type intended to make crofters less economically dependent on the increasingly precarious returns to be got from agriculture.

'The important thing about crofting,' said the Scottish Crofters Union's principal founder, Angus Macleod, in the course of a 1987 television documentary on future prospects for rural Scotland, 'is that it has kept people on the land while farming, on the other hand, has been emptying the countryside.'[8] To appreciate the force of the SCU honorary president's remark, it is necessary only to compare the condition of crofting localities with those of comparable areas where the economic tendencies so persistently favoured by the United Kingdom's agricultural policy-makers were long ago given full rein.

One such area is Morvern – cleared with particular ruthlessness by its early nineteenth-century owners. This was 'one of the most melancholy districts in the whole country,' observed the prominent land reformer John Murdoch in 1879. 'You may walk a distance of nearly 20 miles from the head of Loch Sunart to the head of Loch Aline without meeting a house that is not either a shepherd's or a gamekeeper's.'[9] Since Morvern neither benefited from land settlement nor avoided the impact of the further depopulation which affected so much of the West Highland mainland in the 50 years following the First World War, all that has changed since Murdoch came this way is that shepherds and gamekeepers, too, are now a lot less numerous than they were.

Travel through Morvern or across much of Mull, drive along the Strath of Kildonan in Sutherland or walk to the head of Glen Lyon in Perthshire, and – in all those places where 'viable farms' and 'economic units' have existed for 100 years or more – you will find few people;

see little in the way of habitation; observe practically no recently constructed homes.

But take the road from Barvas to Point of Ness on the Atlantic coast of Lewis, where the weather is far worse and the land generally poorer than in Morvern, Mull, Kildonan or Glen Lyon, but where the average holding extends to four or five acres rather than to the many thousands of acres common in these other locations, and you pass not just the occasional modern house but dozens, even hundreds. Although the north-western corner of Lewis is especially striking in this regard, being the most densely populated part of the British countryside outside the semi-suburban south of England, something of the same contrast between the crofting and non-crofting parts of the Highlands and Islands can be observed very widely.

In this contrast between comparatively thickly-peopled crofting localities and the almost empty glens and straths where crofting long ago ceased to exist there can easily be discerned the makings of the social catastrophe which would have occurred if organisations like the Highland Land League and the Federation of Crofters Unions had not succeeded in standing out against those who wished farms to take the place of those crofts which had survived the clearances.

The concept of the full-time holding, it should be remembered in this context, is inherently elusive. Farms of the sort which appeared perfectly viable to the Napier Commission or the Congested Districts Board in the 1880s and 1890s would have seemed unacceptably small to the Department of Agriculture or the Crofters Commission in the 1950s and 1960s. What was then considered reasonable by agricultural economists would, in turn, be thought quite unacceptable by the prospective full-time agriculturalist of the 1990s – the minimum stocking level required to produce a financially viable farm in most parts of the Highlands and Islands having more than doubled in the interim.

In present circumstances it would take not two or three but scores or even hundreds of Lewis crofts to make a single worthwhile agricultural business. If the previous policy-makers who so badly wanted to move crofting in that direction had been permitted to prevail, there would have been no logical stopping place short of carving all of Lewis, where there are still several thousand smallholdings, into not much more than 20 or 30 large hill farms – farms which, because of the discouraging state of the United Kingdom livestock industry, would themselves today be facing most uncertain prospects.

A crofting policy geared to the creation of viable agricultural units, it can thus be seen, would have had the inevitable effect of directing crofting localities down the path leading to the massive depopulation experienced in recent decades in Scotland's hill farming districts, such as upland Perthshire, where holding after holding has had to be thrown together in a ceaseless attempt to keep essentially marginal agricultural enterprises in profit. Although hill farming on that scale generates both income and employment, its future potential – in terms of its possible contribution to the creation of those diversified rural economies to which the European Community is increasingly committed – is far less than that of crofting.

Crofting already supports more people to the acre, so to speak, than any other non-urban land use currently available in Scotland; more than hill farming; more than commercial forestry of the type which expanded so massively in the course of the 1970s and 1980s; more than deer forest or grouse moor. Where the occupants of those part-time holdings which remain characteristic of most crofting localities have access to additional sources of income, then crofting can readily constitute the basis of a very satisfactory way of living.

Places such as Barra, Assynt and Ardnamurchan – which, in the 1950s and early 1960s, were widely thought to be on the verge of social disintegration – are today in a comparatively flourishing condition. Their economies, while still fragile, have revived considerably. More of their young people can find local employment. Even their total populations, while still far below the levels of a century ago, have begun again to rise.

There is no universal prescription for rural regeneration of this type. Factors as diverse as the growth of fish farming and the expansion of tourism have played a part. But crofting, if not an essential element in the recent success of those parts of the Highlands and Islands which it was once believed to be dragging down, has undeniably made a considerable contribution to their development. By permitting families to have relatively ready access to land and homes – both of which tend to be much less obtainable in those country districts where there are no crofts – it has enabled substantial numbers of people to establish themselves in ways which would have been impossible elsewhere. And though demand for crofts has tended to exceed supply – with the result, which would have been scarcely conceivable in localities like Assynt some 30 or 40 years ago, that further economic growth is actually being inhibited by a lack of affordable holdings – there is certainly no good reason now for continuing to cling to the notion that the fragmentation of landholding is automatically to be deemed a bar to progress.

It is in this sense, as men like James Shaw Grant and Alasdair Fraser first argued in the early 1960s, that the crofting system of the Highlands and Islands, just like its various counterparts elsewhere in Europe, can legitimately be presented – for all the criticisms traditionally hurled in its direction by those who have sought to find essentially agricultural answers to the economic problems of the north of Scotland's crofting areas – as a perfectly rational way of organising the occupation of land in a part of Britain where the returns on farming are, in any event, necessarily limited severely by unalterable climatic conditions.

Grant and Fraser, of course, were making their claims for crofting at a time when there was very little in the way of proof that crofting was, in fact, capable of developing along the lines they were advocating; at a time, moreover, when agricultural policy was explicitly framed in such a way as to accelerate the demise of the part-time farmer. In this respect at least, the Scottish Crofters Union, on attempting to adapt and expand the Grant and Fraser thesis, found things altogether easier in the very different setting of the 1980s.

The emergence of the SCU coincided with mounting criticism of the principles underpinning those rural strategies which had so obsessively subordinated all other considerations to a never-ending quest for increased agricultural production. The union sought repeatedly to make the point that crofting, for all that it had been so markedly at odds with the overall thrust of rural policy when that policy was intended largely to expand farm output, might reasonably be regarded much more favourably in the context of measures designed to meet the quite different demands being made of government by campaigners for an entirely new approach to the management of the British countryside.

The growing political pressure exerted by those campaigners was fuelled by a number of increasingly widespread concerns. Anxiety as to the sheer cost of sustaining the European Community's Common Agricultural Policy was one of these. Public anger at the extent of the damage being done by intensive agriculture to Britain's natural fabric was another. Both the financial and the environmental case against the larger farmer – who had so conspicuously been having things all his own way politically for the previous half century – began to be highlighted in television documentaries, newspaper articles and publications of every kind.

Among the latter were the especially powerful books produced by the conservationist Marion Shoard and the Conservative MP Sir Richard Body – Shoard focussing on the ecological destructiveness of much of modern farming and Body, who both represented a rural constituency and served as the chairman of the House of Commons committee on agriculture, striving to demonstrate that post-war farming policy had served mainly to divert enormous sums of public money into the pockets of so-called agribusinessmen who, in Body's opinion, had profited quite wrongly at the expense both of the taxpayers whom they had so successfully milked of cash and the smaller family farmers whom they had so universally displaced.

Both Marion Shoard and Richard Body were to speak at Scottish Crofters Union conferences. 'Crofters need to develop a long-term strategy for ensuring that their fellow-citizens in the rest of Britain continue to believe that crofting is a good thing,' said Shoard. It was consequently essential, she continued, for crofters to win the backing of the conservation movement by demonstrating that crofting, in contrast to the larger-scale farming which had become so dominant in other parts of the United Kingdom, was perfectly compatible with caring properly for those landscapes and wildlife habitats to which the wider population was attaching more and more importance.[10]

Body, too, stressed the need for crofters to make alliances. The 'rich and large farmers' who had done so well out of government had time, energy and money to devote to the political lobbying at which they had traditionally excelled, said the Conservative MP. Crofters were much less fortunately placed in that respect. Although crofters were beginning to 'make themselves heard' by means of the SCU, it was clearly in their interest to operate increasingly in concert with like-minded groups elsewhere – groups such as the Small Farmers Association which Body had himself helped to establish.[11]

The SCU was to act on that advice. Contact was made with organisations representing smaller farmers throughout the British Isles. By 1990 the SCU – in collaboration with the Small Farmers Association, the Farmers Union of Wales, the Northern Ireland Agricultural Producers Association and the Irish Creamery Milk Suppliers Association – was taking steps to establish a Family Farming Federation capable of exerting meaningful influence on policy-makers in Brussels, London and Dublin.

Nor was the SCU unmindful of Marion Shoard's contention that the case for crofting could be well served, at a time when the political influence of conservationists was so very visibly expanding and at a time, too, when those same conservationists were repeatedly stressing the need to safeguard the natural environment of the Highlands and Islands, by highlighting the many ways in which that environment had been traditionally sustained and protected by crofters.

Within a month or two of the SCU's launch in 1986, the union's first president, Frank Rennie, himself an environmental scientist as well as a crofter, was remarking that the 'great variety of wildlife in crofting areas' was due, in large part, to 'the small-scale, low-input style of agriculture' in which crofters had habitually engaged.[12]

High technology farming of the type common in more southerly and more easterly parts of the United Kingdom, Frank Rennie continued, could only take place profitably on the enormous, prairie-like, holdings which had become typical of localities like East Anglia. There 'massive cereal-producing farms' had 'sacrificed hedges, woods, walls and fences in order to expand'. Birds, animals, flowers and insects had become every bit as expendable as the many families whose smallholdings had been incorporated into the great swathes of territory now ploughed and harvested by huge machines. 'The result,' said Frank Rennie, 'is that both people and wildlife have been driven out of the countryside.'

Such practices, the SCU was to comment over and over again in the years ahead, ought to be contrasted with those prevailing in places like the Western Isles. There the spectacularly flower-rich, bird-rich habitats of the Uist and Barra machair, the union pointed out, had been kept in being as a result of crofters having stubbornly persisted with those traditional land management techniques which most farmers elsewhere had long ago abandoned. It was indicative of growing governmental susceptibility to such arguments that in 1988 the Atlantic coast of Uist and Barra was designated the north of Scotland's first Environmentally Sensitive Area – with several hundred crofters whose holdings

impinged on the machair becoming eligible for substantial cash payments if they agreed to work their land, in effect, without recourse to
the ecologically suspect pesticides, herbicides and other chemicals
applied so widely by those mainland farmers whom crofters had earlier
been urged to emulate.

Environmentalist insensitivity both to crofting aspirations and to the
crofting past – an insensitivity typified by the habitual application of
the term 'wilderness' to many of those parts of the Highland mainland
which were artificially emptied of their populations by nineteenth-
century landlords and which many crofters still feel should one day be
inhabited again – was to result, during the 1980s, in a number of well-
publicised clashes of opinion between crofters and those conservation
groups which were seeking more and more to influence the direction
of public policy in the north of Scotland.

It was sometimes highly tempting to conclude, in such circumstances,
that the undoubted political clout exercised by organisations like the
Royal Society for the Protection of Birds – which, by 1990, had many
more members than there were people living in the entire area subject
to the jurisdiction of the Highlands and Islands Development Board –
would forever be deployed in opposition to crofting interests. But the
undoubted success of the Uist and Barra ESA – from which some 200
island crofters were gaining financially within a year of the scheme's
commencement – pointed in the altogether more hopeful direction
signposted by Marion Shoard; as did the fact that the RSPB itself
was increasingly advocating EC Common Agricultural Policy reforms
of a type which would result in a greater proportion of CAP expenditure being allocated to smaller-scale livestock producers of the crofting
type.

In aspiring to do rather better out of the Common Agricultural Policy,
the crofting population might seem simply greedy; for it tends to be
assumed widely, as noted in this book's opening pages, that crofters are
already profiting hugely from a quite unrivalled range of publicly
financed subsidies. In fact, however, this is not at all the case. Many
previously available subventions and concessions were withdrawn or
cancelled in the 1970s and 1980s. Much of what remains in the way
of public expenditure on crofting is a lot less generous than is frequently
asserted.

Lime subsidies ended in 1976. Cropping grant was abolished in 1981. The township road scheme was terminated in 1985. Charges have been introduced for agricultural advisory services which were previously free. Most significantly of all, the introduction of the poll tax in 1989, as already mentioned, was unaccompanied by any crofting concession corresponding to the 50 per cent derating previously applied to croft homes.

There remain, then, only two significant forms of government-funded financial assistance which are peculiar to crofters. One consists of those crofter housing grants and loans administered by the Department of Agriculture. The other is the Crofting Counties Agricultural Grants Scheme which is the responsibility of the Crofters Commission.

Both the housing scheme and CCAGS cost the UK taxpayer upwards of £2 million annually. In relation to overall public expenditure, these are not large sums. It can even be argued that one of them, the amount devoted to crofter housing, can actually be reckoned a net gain to the Exchequer – on the basis that, as was noted in the previous chapter, it would be more costly to meet the legitimate housing needs of crofting communities by the standard means of assisting local authorities or similar bodies to construct homes for rent. While no similar case can be made on behalf of CCAGS, it should perhaps be emphasised that total annual spending under that heading is of much the same order of magnitude as that involved in the provision of perhaps a quarter mile of urban motorway.

A crofter who receives a grant from the Crofters Commission to assist with the erection of a fence or the construction of a shed is obliged, of course, to make a substantial personal investment in the project in question. The same conditions applied to the monies made available to participants in the various development programmes initiated in the Highlands and Islands in the course of the 1980s. The fact that these programmes have resulted in many new buildings and other permanent improvements consequently testifies every bit as much to the skill, enterprise and commitment of the crofting community as it does to the generosity of the various funding bodies.

That generosity should be appreciated. But it should not be exaggerated. Measures of the IDP and ADP type have clearly been of immense help to crofters. But that help has been of an essentially temporary character. And crofters, who failed so signally to profit from the much

213

more permanent agricultural support structures put in place by post-war UK governments, have not done significantly better as a result of the modifications made to those structures in the wake of Britain's entry into the European Community in 1973.

Public spending on farming, though now increasingly within the jurisdiction of the European Commission, most certainly did not decrease after 1973 – the older hill sheep and hill cattle subsidies, for example, simply being replaced by still more generous disbursements. Crofters, of course, received some share of these as a result of their entitlement to hill livestock compensatory allowances, sheep annual premium, suckler cow premium and the various other income aids by which the EC sought to protect and bolster agricultural producers.

But crofters have not been among the major beneficiaries of the EC's Common Agricultural Policy. The SCU, as already mentioned, had to press hard to get agreement that suckler cow premium, for instance, should be obtainable by part-time crofters. Also, hill livestock compensatory allowances and sheep annual premium, because both – with the minor exception of the compensatory allowance increment introduced in the HIDB area in 1984 – tended from the start to take the form of flat-rate headage payments on the animals concerned, have generally gone overwhelmingly to those farmers with the biggest flocks and herds.

The almost unbelievable scale of the resulting discrepancies can readily be illustrated by calculating the support payments due, in 1988, to two very disparate sheep producers in two equally distinctive parts of the north of Scotland. One is a crofter with 50 ewes which he keeps in a very hard, wet, exposed part of Lewis. The other maintains 3,000 ewes on his extensive estate in a comparatively fertile and sheltered mainland strath. The crofter, in the course of the year in question, received under £700 in the shape of sheep annual premium and hill livestock compensatory allowance cheques. The landlord collected over £40,000.

Even allowing for the fact that the estate owner, unlike the crofter, may have been employing labour, the extent of that disparity is quite remarkable. Yet this tendency to favour the larger farmer on better land, at the expense of the smaller farmer on poorer land, has been an enduring feature of the Common Agricultural Policy.

It was bad enough that the CAP, as applied by British governments to the Highlands and Islands at any rate, took no account of the circumstances which resulted in the typical crofter tending to occupy less productive land than his farming competitor, to be more remote from both markets and suppliers – and, of course, to be much less able than the bigger operator to achieve any worthwhile economies of scale. It was even worse that overall spending on measures of the hill livestock compensatory type has generally accounted for under three per cent of total CAP expenditure.

The remaining 97 per cent of a budget – which has itself at times swallowed up some 75 per cent of all the funds at the EC's disposal – has been allocated to the CAP's guarantee sector. This is primarily concerned with commodity price support. Much of it has been spent on keeping up the cost of cereals.

Thus the overwhelming bulk of CAP expenditure, as observed pointedly by agricultural economist John Bryden, who once headed the Highland Board's Land Division and who subsequently became programme director of the Arkleton Trust rural research organisation, has gone 'to those farmers who produce most, or, more precisely, who produce most of the commodities receiving most support. By and large, these are definitely not small farmers in marginal areas.' They are, in fact, the larger farmers occupying the EC's most productive regions – places like East Anglia and the northern part of France. It is farmers of that sort, not crofters or the continental counterparts of crofters, who have indubitably gained most from the CAP – some 80 per cent of EC spending on agriculture, it has been calculated, having gone to a mere 20 per cent of the Community's more prosperous agricultural producers.[13]

To Sir Richard Body, addressing delegates to the SCU's 1987 annual conference, there seemed to be a simple explanation for the United Kingdom's failure to follow the example of many of its EC partners and impose upper limits on an individual landholder's eligibility to claim the various allowances and premiums available to agriculturalists. 'Why is it that the rules are so different in this country? Why is it that some farmers are making as much as £50,000 a year from hill livestock compensatory allowances and other payments? It is because it is the voice of the bigger farmer that is heard and heeded by our Government.'[14]

But slowly that was changing. The SCU, by itself, could not persuade politicians to engage in a radical readjustment of policy. And though it was possible for the union to bring rather more political influence to

215

bear as a result of its steadily developing links with broadly similar groups elsewhere, it was clear that the really irresistible pressure for change would emanate from other sources; from those countries like the United States and Australia which, in order to gain freer access to European markets, were increasingly insisting on some dismantling of CAP tariffs; from EC consumers pressing for cheaper food and lower taxes; from conservation groups calling more and more stridently for agricultural policies to be given a much greater environmental dimension.

By the winter of 1990–91, when international talks intended to reinforce the General Agreement on Tariffs and Trade threatened to break down on the CAP issue, far-reaching reform was beginning to appear inevitable. Ceilings were already being imposed on headage payment schemes. EC agriculture commissioner Ray McSharry was making clear his own preference for channelling CAP aid in the direction of Europe's smaller farmers.

In a purely British context, of course, this was to threaten a complete overturning of the principles which had shaped agricultural policy for 50 years. There was much talk, as there had been since CAP restructuring was first seriously mooted, of the need to sustain the country's 'more efficient' – meaning more substantial – farmers. But the McSharry approach had its UK supporters also – most significantly among the conservationists.

'A criticism which is levelled at discrimination in favour of small-scale farmers,' the RSPB had commented in 1988, 'is that it ignores the benefits to consumers and taxpayers of having production concentrated on the most efficient farms. However, it is in the areas of just such "efficient" production that the costs in terms of wildlife and the environment, rural employment and public resources have been greatest. The need to reform the CAP provides an opportunity to think about efficiency in a less narrow way.'[15]

With previously heretical thinking of this type showing some signs, by the early 1990s, of becoming itself the new orthodoxy, there was every possibility that crofting – paradoxically – would seem a lot less anachronistic in the twenty-first century than it had done for much of the period examined in this book. Although there were inevitable anxieties among crofters as to how they might be affected by the many policy changes likely to be forthcoming, it seemed to SCU director George Campbell – who was able to discuss crofting prospects directly with EC commissioner McSharry in February 1991 – that the general thrust of what was being proposed would not be, by any means, calamitous from the crofting point of view.

The crofting areas of the Highlands and Islands continue to experience economic and social difficulties. For all that so many new houses have been built in crofting communities in recent years, it is the case – as was acknowledged in 1990 by the government housing agency Scottish Homes – that those parts of Scotland with the highest proportion of dwellings still lacking inside sanitation are to be found not, as one might loosely suppose, in inner city districts of the Central Belt but in places like the Western Isles.

Croft homes of that sort are almost invariably occupied by elderly people who lack sufficient cash to take advantage of crofter housing scheme finance – which the Department of Agriculture only makes available to those individuals with the means to make the necessary loan repayments. It is consequently among the older age groups that crofting status remains most closely linked with the poverty and disadvantage which were once so characteristic of the crofting population as a whole.

Comparatively few crofters, of course, are enormously well-off. Crofting income surveys commissioned from the North of Scotland College of Agriculture by the SCU in the later 1980s demonstrated that the average crofter was then earning not much more than £8,000 annually from all sources – with returns on agricultural activities being no more than a few pence an hour in many cases. Although it has been one of this book's consistent themes that crofters ought not to be confused with farmers, the steeply falling livestock prices of 1989 and subsequent years have undoubtedly set back the prospects of many of those younger people who – in one of the more encouraging developments of recent times in the Highlands and Islands – are again becoming actively involved in crofting.

Both young men and young women, as one might expect, generally responded much more enthusiastically than their elders to the various incentives which the IDP, the ADP and related developmental initiatives made available to those willing to engage in expanding livestock production in the crofting areas. The results of their endeavours are everywhere to be seen.

It is frequently still asserted, of course, that crofting is not what it was; that the land was more assiduously worked and better managed in the past; that too many townships are unkempt and unattractive in appearance; that holdings are uncropped and hill pastures neglected. It would be less than honest to maintain that there is not a grain of truth in such assertions.

But those most given to bemoaning, like so many latter day Alasdair Alpin MacGregors, the absence of the house cow from the byre and the oat stook from the field, are inclined to overlook the extent to which wider economic trends – of a kind that were visibly changing the nature of crofting as far back as the 1930s – have now wholly altered the pattern and purpose of the crofter's production from his holding.

The many younger and more active crofters to be found in the Highlands and Islands of the 1990s are not seeking simply to provide themselves and their families with milk, mutton and potatoes. They are producing high quality lambs and calves for highly discriminating national and international markets. They ought to get a little bit more credit than they do for what they have accomplished.

Younger people, as observed already, have been extremely prominent in the running of the SCU. Although crofters in their teens, their twenties, their thirties or their forties are still less numerous than crofters in their sixties or their seventies, the fact that younger age groups are now well represented in a growing number of crofting localities is a powerful indication that those depopulating processes which featured so largely in this book's earlier chapters have been brought widely to an end.

Where once there were large numbers of vacant and abandoned crofts there is today intense demand for land. The SCU has found it very difficult to discover a satisfactory means of tackling the problems confronting those younger people from a Highlands and Islands background who are being outbid in the mounting competition for such holdings as become available, but the current upsurge of interest in obtaining access to crofting is a much more encouraging phenomenon than the seemingly unstoppable exodus which preceded it.

'In island after island and place after place,' said Western Isles MP Malcolm MacMillan of the crofting areas in 1955, 'the population is dying away.' Throughout the north of Scotland, the Highland Panel observed at the same time, there was an endemic sense of 'pessimism

and defeatism'. Young people were leaving 'less because of any imminent financial problems than because they had a feeling that there was
no chance of their getting on if they stayed'. The crofting life, it was
generally agreed, had nothing very much to recommend it.[16]

'Our survey of the crofting system,' members of the Taylor Commission reported in 1954, 'leaves one dominant impression in our minds.
It is a system which, as now organised, is fighting a losing battle against
the social and economic forces of the day. In many places . . . it is in
a state of decline and in some, indeed, of dissolution.'[17]

No modern analysis of crofting circumstances could possibly come to
so unrelievedly gloomy a conclusion. Not only do the external forces
to which the Taylor Commission referred – prominent among them,
of course, the impetus towards the enforced amalgamation of croft holdings – no longer look quite so all-conquering as once they did. It is
increasingly feasible to maintain, as immediately preceding pages have
sought to demonstrate, that the traditional crofting structure is ideally
suited to capitalise on the opportunities presented by UK and EC rural
policies shaped more and more by wholly new determinants.

That is one factor in the confidence which so many crofters have
today in crofting. 'For far too long,' SCU president Frank Rennie told
a union conference in 1988, 'crofting has been portrayed as something
that's on the way out. And crofters have been presented to the world
as a pretty incompetent bunch of folk eternally in search of the next
hand-out.' Crofters had never actually conformed to that unflattering
stereotype, Frank Rennie said. They had no intention of conforming
to it now. 'We are independent people,' the SCU president insisted.
'We are proud of what we do. We are convinced that crofting can have
a first-rate future.'[18]

Anyone who doubted the crofting community's capabilities, Frank
Rennie concluded on that occasion, need only visit the Moidart home
of David and Georgina Duncan, joint winners of the Crofter of the Year
award which the SCU – with the financial support of the Royal Bank
of Scotland – had instituted some months earlier.

In the space of about four years the Duncans, operating as a team and
turning out in every kind of weather, had drained, rehabilitated and
reseeded acre after acre of badly run-down land. They had built a cattle
shed and a sheep shed. They had diversified into the letting of holiday
caravans. They had improved their holding's stock to the point at

which their lambs and calves regularly commanded premium prices at auction. And they had erected, entirely unaided, some 5,000 yards of high grade fencing.

'We have worked hard,' David Duncan told a member of the Crofter of the Year judging panel. 'But now we can see the results. And that does something for you.'[19] It was a comment which went no small way to summing up the state of crofting in the decade which had seen so much effort expended – in the end so fruitfully – on the creation of the SCU.

Crofting has long been on the defensive; so long that it is sometimes hard to grasp that things have finally begun to change for the better; so long that it is difficult to appreciate the extent to which the wider world in which the crofting case has got to be advanced politically is now capable of being persuaded that crofting, or something very like it, makes more and more sense in the context of contemporary environmental and developmental priorities; so long that crofters are themselves occasionally reluctant to believe that it is beginning to be possible to be optimistic about crofting.

Circumstances have seldom made it easy for crofters to be properly self-confident. At the end of one of the many tours he made of the Highlands and Islands in the 1870s, John Murdoch wrote:

> We have to record the terrible fact that from some cause or other, a craven, cowed, snivelling population has taken the place of the men of former days. In Lewis, in the Uists, in Barra, in Skye, in Islay, in Applecross and so forth, the great body of people seem to be penetrated by fear. There is one great, dark cloud hanging over them in which there seem to be the terrible forms of devouring landlords, tormenting factors and ubiquitous ground-officers. People complain; but it is under their breaths and under such a feeling of depression that the complaint is never meant to reach the ear of landlord or factor. We ask for particulars, we take out a notebook to record the facts; but this strikes a deeper terror. 'For any sake do not mention what I say to you,' says the complainer. 'Why?' we naturally ask. 'Because the factor might blame me for it.'[20]

The younger crofters whom Murdoch successfully urged to rebel were to go some way to combating those apprehensions. 'They must put the fear of the landlord and his satellites . . . out of the old men,' one lad from Lochcarron told a Highland Land League meeting in 1886. 'They must try and make them realise that the powers that be are nothing to the powers that the generality of the people possess.'[21]

But for all that the Land League was itself instrumental in eliminating the once constant threat of enforced dispossession, the crofting community was left to struggle – often it seemed unavailingly – with other more intangible, more insidious but nonetheless potent assaults on its collective self-esteem.

The lack of economic opportunity at home was the basic underlying cause of the depopulation which so stripped the Highlands and Islands of those energies and talents which were to be put so productively at the disposal of so many other societies elsewhere. But depopulation was facilitated, too, by the ceaseless devaluing of the crofting community's cultural heritage; by the conviction, on the part of almost all those in authority that practically everything about the crofting way of life, starting with the Gaelic language, was intrinsically inferior; by the practically universal assumption that the path to betterment must forever point outward from the croft.

'The language and lore of the Highlanders being treated with despite,' observed John Murdoch, 'has tended to crush their self-respect and to repress that self-reliance without which no people can advance.' Crofters, Murdoch believed, had to be made to feel that 'they themselves, and the things which belong to them, are of greater value in the world than they have for some time been taught to regard them'. Otherwise they would continue timorous, fearful and submissive – endlessly 'afraid to open their mouths'.[22]

But that was not to be readily accomplished. To Frank Fraser Darling, writing in the 1940s, there was nothing more indicative of what needed to be put right in the Highlands and Islands than the 'lassitude' and 'depression' then so characteristic of so many crofting townships. It was not that crofters lacked 'innate ability', Fraser Darling continued. It was that no worthwhile attempt had been made 'to set free this real ability in the crofting areas and create an active and healthy social organism'.[23]

Wherever it was that boys and girls from a crofting background might advance and prosper, it was generally considered, it would not be in the places where they had been born and brought up. And so, as Marion Fraser Darling was to tell the Taylor Commission in the early 1950s, those young people who chose to remain in their own localities were automatically deemed less successful, possibly even by themselves, than those of their contemporaries who had moved south or even left the country altogether. The son who stayed at home to work the family croft was not unappreciated by his parents. But the other son, whose graduation photograph was prominently displayed on the kitchen dresser and who was typically said to be 'doing very well' in Glasgow or Vancouver, was frequently a greater source of family pride.

Initiative and enterprise were not qualities thought missing in those Highlanders and Hebrideans who contributed so massively to the development of nations like the United States and Canada. But they did seem strangely lacking in the places from which these most success-ful emigrants had come.

'There is a feeling among Lewis crofters,' it was noted by a group of islanders in 1944, 'that they prefer to have a landlord before whom they can lay their difficulties and disputes and who will attend to their wants; an authority that relieves them of responsibility.'[24]

Where the landlord fell down in that respect, there was no scarcity of public bodies to step into the breach – to the extent that crofting townships seemed at times to be incapable of taking the simplest of decisions without reference to some official agency. 'The people had come to rely to a certain extent upon direction from superior bodies,' runs the Taylor Commission's record of the evidence presented to its members by Roddy MacFarquhar of the Scottish Agricultural Organisa-tion Society, 'and he felt it essential in any future planning for the High-lands that a general community spirit should be redeveloped and responsibility given as far as possible to local bodies.'[25]

But the Taylor enquiry team were to pay little heed to such pleas. Instead they were to recommend the establishment of yet another administrative organisation, the Crofters Commission. 'Presumably the proposed authority,' remarked the economist Alex Cairncross in the course of the Taylor Commission's internal debates about the new body's remit and functions, 'would not have crofters on it.' From that particular proposition there was no dissent. The one group of people who ought not to be involved directly in the management of crofting, it had long since been concluded, were those who had chosen to make their lives on crofts.[26]

Crofting communities should be preserved, the Taylor Commission concluded, 'because they embody a free and independent way of life which in a civilisation predominantly urban and industrial in character is worth preserving for its own intrinsic quality'.[27] Up to a point, that was to advance the claim of crofting in the sense in which it has been stated in this book. But only up to a point; for what the Taylor Commission – perhaps inevitably in the circumstances of the early 1950s – felt bound, in effect, to insist upon was something which the author of this book – written from the very different perspective of the 1990s – would most determinedly deny: that crofters are incapable of managing their own affairs.

For the greater part of this century, as this book has amply demonstrated, institutions of the sort which the Taylor Commission strove to create have exercised a pervasive influence on the development of crofting. It is time, perhaps, for the role of these extraneous agencies to be curtailed. Not because of any incompetence or lack of motivation on the part of their current staffs. Certainly not because they have made no positive contribution to the overall wellbeing of the Highlands and Islands. But because crofters today – whatever may or may not have been the case in the past – are perfectly capable of taking charge of crofting.

Precisely how such a transfer of power is to be arranged is obviously a matter for debate. It need not involve the complete dismantling of organisations like the Crofters Commission. But it certainly ought to involve some considerable enhancement of the crofting community's capacity to advance its own interests in ways that seem best to it. It is in relation to this pressing requirement to make crofters responsible for the management of crofting that the possibility of croft land passing into community ownership is unreservedly to be welcomed.

That possibility was raised by the Conservative Government in connection with the Department of Agriculture's crofting estates. But as Angus Macleod, for one, quickly hastened to point out, there is no overriding reason why the community ownership principle should not be applied to all croft land. Nor indeed is there any reason why the total area in crofting occupation should not again begin to be expanded – crofting, as previous pages have striven to show, being much more advantageous, from both the social and the environmental point of view, than most of the other uses to which land in the north of Scotland is presently being put.

Towards the end of John McGrath's 1973 play, *The Cheviot, the Stag and the Black, Black Oil,* now recognised to be one of the classic productions of modern Scottish theatre, two of the young Highlanders among the characters remark: 'The people do not own the land. The people do not control the land.' And the play closes with the singing of one of the songs with which Mary MacPherson, *Mairi Mhor nan Oran,* more than a century ago inspired the Highland Land League.

'When I am in my coffin,' *Mairi Mhor* predicted, 'my words will be as a prophecy. And there will return the stock of the tenantry who were driven over the sea. And the gentry will be routed, as they, the crofters, were. Deer and sheep will be wheeled away and the glens will be tilled. There will be a time of sowing and of reaping; a time of reward for the robbers. And the cold, ruined stances of houses will be built on by our kinsmen.'

To be present in one of the many crofting communities where McGrath's play was performed in the early 1970s was to realise anew the sheer power of that vision; to recognise, that for all the heartbreaks and the disappointments which crofters and their families have experienced in the 200 years since the clearances, there is still the will among them to make the land their own. Although the Highlands and Islands future will not be precisely as big Mary MacPherson foresaw it in the 1880s, there are today good grounds to think that both crofters and crofting will do rather better in the century about to start than they did in the one now coming to a close.

REFERENCES

Abbreviations used below are as follows:

CC   Crofters Commission
DAFS   Department of Agriculture and Fisheries for Scotland
HIDB   Highlands and Islands Development Board
HMSO Her Majesty's Stationery Office
SCU   Scottish Crofters Union
SRO   Scottish Record Office
TC   Taylor Commission – Commission of Enquiry into Crofting Conditions
       1952–54
WHFP West Highland Free Press

Introduction
1. Sanderson of Bowden speaking at SCU conference, Liniclate, 16 March 1989.
2. *The Crofter*, June 1989.

Chapter One. Better Times, Worse Times
1. D. Gillies, *Eilean a' Cheo: Annals of Skye, 1945–46*, Nethy Bridge. Unpaginated, photocopied diary.
2. W. A. Hance, *The Outer Hebrides in Relation to Highland Depopulation*, New York, 1949, 43.
3. A. Geddes, *The Isle of Lewis and Harris*, Edinburgh, 1955, 278. TC, Evidence: SRO, AF81/10.
4. F. Fraser Darling, *West Highland Survey*, London, 1955, 356.
5. TC, Evidence: SRO, AF81/10.
6. A. Collier, *The Crofting Problem*, Cambridge, 1953, 99. TC, Evidence: SRO, AF81/15.
7. CC, *Report for 1957*, 15.
8. S. Gordon, *The Immortal Isles*, London, 1926, 155–6.
9. R O'Malley, *One Horse Farm*, London, 1948, 198–9, 207.
10. TC, Evidence: SRO, AF81/7.
11. TC, *Report*, HMSO, 1954, 36.
12. TC, *Report*, 31.
13. TC, Evidence: SRO, AF81/14.
14. C. Addison, *A Policy for British Agriculture*, London, 1939, 14–15, 55.

15 J. A. Symon, *Scottish Farming Past and Present*, Edinburgh, 1959, 223–9. *Report of the Committee on Hill Sheep Farming in Scotland*, HMSO, 1944, 19–20.
16 Collier, *Crofting Problem*, 9.
17 C. MacDonald, *Highland Journey*, London, 1943, 53.
18 Darling, *Survey*, 12.
19 *Glasgow Herald*, 8 November 1946.
20 Hilleary Report, 1938: SRO, AF81/10.
21 TC, Evidence: SRO, AF81/4.
22 D. Keir, 'The Desolation of the Highlands', *Fact*, London, 1938, 6–7.
23 *Hansard*, Lords, 13 July 1939.
24 Scottish Office to Treasury, 10 June, 20 July 1939: SRO, DD15/15.
25 Ramsay to Morrison, 27 April 1936: SRO, DD15/54/1.
26 *Scotsman*, 8 January 1936.
27 *Hansard*, Commons, 1 April 1936.
28 Hilleary Committee Papers, 1936–39: SRO, DD15/60, SEP12/80.
29 *Hansard*, Commons, 16 December 1936. A. F. Cooper, *British Agricultural Policy 1912–1936*, Manchester, 1989, 161.
30 *Glasgow Herald*, 22 December 1936.
31 *Scotsman*, 23 February 1938.
32 Hilleary Report, 1938: SRO, AF81/10.
33 A. MacEwen and J. L. Campbell, *Act Now for the Highlands and Islands*, Edinburgh, 1938, 5–7.
34 Scottish Office memorandum, 2 March 1939: SRO, DD15/14.
35 Scottish Office to Treasury, 10 June 1939: SRO, DD15/15.
36 Scottish Office to Treasury, 14, 20 July 1939: SRO, DD15/15. *Hansard*, Commons, 1 August 1939.
37 MacDonald to Colville, 2 September 1939: SRO, DD15/17.

Chapter Two. No Answer to the Crofting Problem

1 Collier, *Crofting Problem*, 66.
2 P. L. Payne, *The Hydro*, Aberdeen, 1988, 43.
3 Payne, *Hydro*, 74.
4 DAFS, *Report for 1939–48*, 45.
5 TC, Evidence: SRO, AF81/11.
6 TC, Evidence: SRO, AF81/10.
7 *Report by the Scottish Land Settlement Committee*, HMSO, 1945, 52–3. Darling, *Survey*, 35. TC, *Report*, 26.
8 *Hansard*, Commons, 27 January 1955.
9 TC, *Report*, 20.
10 Darling, *Survey*, 100.
11 TC, Evidence: SRO, AF81/17.
12 TC, Evidence: SRO, AF81/14.
13 TC, *Report*, 19. Darling, *Survey*, 38. Highland Panel paper, June 1948: SRO, SEP12/118.
14 Darling, *Survey*, 223.

15 Darling, *Survey*, 219.
16 TC, Evidence: SRO, AF81/10.
17 Darling, *Survey*, 13.
18 TC, *Report*, 38.
19 TC, Evidence: SRO, AF81/8, AF81/11.
20 Darling, *Survey*, 44.
21 Darling, *Survey*, 282.
22 Collier, *Crofting Problem*, 4.
23 TC, Evidence: SRO, AF81/7.
24 Ibid.
25 TC, Evidence: SRO, AF81/8.
26 Ibid.
27 Ibid.
28 Synod of Argyll, April 1946: SRO, DD15/63.
29 Geddes, *Lewis and Harris*, 79. Darling, *Survey*, 253. TC, *Report*, 20.
30 TC, Evidence: SRO, AF81/8.
31 TC, *Report*, 33.
32 *Oban Times*, 22 November 1952.
33 TC, *Report*, 40–41.
34 Scottish Office note to Ross and Cromarty County Council, June 1942: SRO, DD15/63.
35 J. M. Bannerman, 'Post-War Development in the Highlands', *Proceedings of the Royal Philosophical Society of Glasgow*, 1945, 76.
36 K. O. Morgan, *Labour in Power, 1945–51*, Oxford, 1985, 307.
37 TC, Evidence: SRO, AF81/8.
38 TC, *Report*, 36.
39 Scottish Office correspondence, 1951–52: SRO, AF81/28. *Glasgow Herald*, 26, 29 November 1952.
40 Lewis Association, *Report on Agriculture*, Stornoway, undated, 22.
41 DAFS, *Scotland's Marginal Farms: The Highlands*, HMSO, 1947, 42.

Chapter Three. Looking for a New Approach

 1 Scottish Office memorandum, 1 February 1939: SRO, DD15/12.
 2 Highland Panel briefing paper, March 1947: SRO, SEP12/118.
 3 TC, Evidence: SRO, AF81/10.
 4 TC, Evidence: SRO, AF81/7, 10, 12–14, 19.
 5 TC, Evidence: SRO, AF81/7.
 6 Highland Panel minutes, 20 October, 3 November, 19 December 1950: SRO, SEP12/2.
 7 Notes for the Secretary of State, 16 January 1951: SRO, AF81/1.
 8 Note by Secretary of State, 14 November 1950: SRO, AF81/1.
 9 Highland Panel press statement, 19 December 1950: SRO, AF81/1.
10 *Glasgow Herald*, 19 January 1951. *An Gaidheal*, May 1951.
11 *Glasgow Herald*, 1 November 1950. *Scotsman*, 20 December 1950.
12 *Stornoway Gazette*, 11 January 1952.

13 TC, Evidence: SRO, AF81/9.
14 TC, *Report*, 9.
15 TC, Note on Shetland visit: SRO, AF81/15.
16 TC, Note on Western Isles visit: SRO, AF81/16.
17 TC, Evidence: SRO, AF81/3, 11.
18 TC, Evidence: SRO, AF81/17.
19 TC, Evidence: SRO, AF81/8.
20 TC, Evidence: SRO, AF81/2, 21.
21 TC, Evidence: SRO, AF81/22.
22 TC, Evidence: SRO, AF81/23.
23 TC, Evidence: SRO, AF81/22.
24 This comment and the following from TC, *Report*, 8–9, 21–4, 27–8, 40, 44, 80–1, 87.
25 *Hansard*, Commons, 27 April 1954.
26 Notes for Minister, April 1954: SRO, AF81/167.
27 Note by DAFS, 26 February 1954: SRO, AF81/167.
28 Memo by Secretary of State, 28 August 1954: SRO, AF81/167.
29 This comment and the following from *Hansard*, Commons, 27 January 1955.
30 A. Cairncross, 'Foreword' in Collier, *Crofting Problem*, ix.
31 Collier, *Crofting Problem*, 1–2.
32 F. Fraser Darling, *Island Farm*, London, 1943, 197.
33 Darling, *Survey*, 192.
34 Darling, *Survey*, 193–4, 359–60, 362–3.
35 J. M. Boyd, *Fraser Darling's Islands*, Edinburgh, 1986, 207.
36 Boyd, *Islands*, 209.
37 F. Fraser Darling, 'Ecology of Land Use' in D. C. Thomson and I. Grimble (eds), *The Future of the Highlands*, London, 1968, 53–4.

Chapter Four. False Start for the Crofters Commission

1 CC, *First Report*, 7.
2 Ibid.
3 CC, *First Report*, 10.
4 *Skye Clarion*, June 1956.
5 CC, *Report for 1960*, 11.
6 CC, *Report for 1960*, 10.
7 CC, *Report for 1959*, 8; *Report for 1960*, 10.
8 CC, *Report for 1960*, 9.
9 *Skye Clarion*, November 1956.
10 CC, *First Report*, 5, 23. *Skye Clarion*, November 1956.
11 *Skye Clarion*, November 1956.
12 CC, *Report for 1959*, 8.
13 CC, *First Report*, 16; *Report for 1957*, 5.
14 CC, *First Report*, 6.
15 *Skye Clarion*, May 1956.

16 *Skye Clarion*, April 1956. Urquhart, confidential memorandum, 16 April 1959: SRO, AF81/150.

17 CC, *Report for 1959*, Appendix IX.

18 CC, *Report for 1958*, 9.

19 Urquhart, confidential memorandum, 16 April 1959: SRO, AF81/150.

20 CC, *Report for 1959*, 16–23.

21 *The Scottish Economy 1965 to 1970*, HMSO, 1976, 28. G. Clark, 'Farm Amalgamations in Scotland', *Scottish Geographical Magazine*, 1979, 94.

22 TC, *Report*, 30.

23 F. Fraser Darling, *Crofting Agriculture*, Edinburgh, 1945, 1.

24 Darling, *Survey*, 203.

25 CC, *First Report*, 22; *Report for 1960*, 7.

26 *Report of the Commissioners of Inquiry into the Conditions of Crofters and Cottars*, HMSO, 1884, 39.

27 *Oban Times*, 11 February 1939. The Mallaig resolutions are preserved in SRO, AF81/8.

28 Darling, *Survey*, 283.

29 L. Leneman, *Fit for Heroes: Land Settlement in Scotland after World War I*, Aberdeen, 1989, 31.

30 TC, Evidence: SRO, AF81/7.

31 *Stornoway Gazette*, 30 August 1960.

32 *Stornoway Gazette*, 6 September, 18 October 1960.

33 *Stornoway Gazette*, 6, 13 December 1960.

34 *Hansard*, Commons, 27 January 1955.

35 *Hansard*, Commons, 15 March, 20 April 1961.

36 Quoted: *Stornoway Gazette*, 22 August 1961.

Chapter Five. The Commission Tries Again

1 CC, *Report for 1962*, 11.

2 CC, *Report for 1962*, 8.

3 *Stornoway Gazette*, 28 September 1963.

4 CC, *Report for 1963*, 22.

5 CC, *Report for 1963*, 26.

6 CC, *Report for 1964*, 30–1.

7 CC, *Report for 1966*, 11.

8 *Stornoway Gazette*, 12 March 1966.

9 CC, *Report for 1963*, 28.

10 *Stornoway Gazette*, 25 June 1966, 19 August 1967.

11 CC, *Report for 1966*, 11; *Report for 1967*, 8–9; *Report for 1973*, 1. *Stornoway Gazette*, 30 November 1968.

12 CC, *Report for 1968*, 27–38.

13 TC, Note on Skye visit: SRO, AF81/18.

14 Darling, *Survey*, Preface, viii.

15 *People's Journal*, 9 November 1968.

16 *Highland News*, 24 October 1968.

17 *Scotsman*, 24 October 1968.
18 *Daily Record*, 24 October 1968. *Glasgow Herald*, 30 October 1968.
19 *Times*, 24 October 1968. *Scotsman*, 24 October 1968.
20 *Economist*, 26 October 1968. *Inverness Courier*, 25 October 1968.
21 *People's Journal*, 9 November 1968. *Inverness Courier*, 28 January 1969. *Scotsman*, 13 February 1969. *Stornoway Gazette*, 15 February 1969.
22 *Oban Times*, 10 April 1969. *Glasgow Herald*, 12 June 1969.
23 *Press and Journal*, 20 August 1969.
24 *Stornoway Gazette*, 19 July 1969.
25 CC, *Report for 1969*, 7. *Inverness Courier*, 21 November 1969.
26 *Scotsman*, 5 August 1972. WHFP, 28 July, 15 September 1972.
27 *Press and Journal*, 7 August 1972. *Northern Times*, 11 August 1972.
28 *Scotsman*, 29 March 1969. WHFP, 11 August 1972.
29 *Glasgow Herald*, 8, 12 July 1969.
30 WHFP, 24 November 1972, 19 October 1973.
31 *Inverness Courier*, 16 March 1973. WHFP, 29 June 1973.
32 *Hansard*, Commons, 22 January 1974.
33 WHFP, 28 March, 6 June, 25 July 1975.
34 *Hansard*, Scottish Grand Committee, 16 December 1975.
35 *Hansard*, Commons, 2 March 1976. CC, *Report for 1976*, 1.
36 *Highland News*, 15 January 1970.
37 *Stornoway Gazette*, 31 January 1970, 19 September 1975.
38 WHFP, 27 August 1976.
39 CC, *Report for 1981*, 2. *Press and Journal*, 9 November 1964. WHFP, 16 November 1984.

Chapter Six. The Highland Board Tackles the Land Question
1 D. Simpson, 'Investment, Employment and Government Expenditure in the Highlands', *Scottish Journal of Political Economy*, 1963, 271.
2 F. R. Hart and J. B. Pick, *Neil M. Gunn*, London, 1981, 234.
3 *Hansard*, Commons, 16 March 1965.
4 Boyd, *Islands*, 1986, 197. Darling, 'Ecology of Land Use', 1968, 51.
5 A *Programme of Highland Development*, HMSO, 1950, 14.
6 TC, Evidence: SRO, AF81/167. TC, *Report*, 37.
7 CC, *Report for 1960*, 18. *Stornoway Gazette*, 18 October 1960.
8 Advisory Panel on the Highlands and Islands, *Land Use in the Highlands and Islands*, HMSO, 1964, 1, 23, 26, 28.
9 HIDB, *First Report*, 5.
10 J. Grassie, *Highland Experiment*, Aberdeen, 1983, 5.
11 Grassie, *Experiment*, 45–6.
12 *Hansard*, Commons, 17 March 1965.
13 HIDB, *Report for 1968*, 31.
14 F. Gillanders, 'Economic Life' in Thomson and Grimble, *Future of the Highlands*, 128.
15 HIDB, *First Report*, 29.

16 HIDB, *First Report*, 3; *Report for 1967*, 23; *Report for 1968*, 13.

17 HIDB, *First Report*, 29.

18 Grassie, *Experiment*, 77.

19 HIDB, *Report for 1968*, 59. HIDB, *Strath of Kildonan: Proposals for Development*, 1970, 10.

20 HIDB, *Kildonan*, 16, 30.

21 Darling, *Survey*, 47. TC, Evidence: SRO, AF81/8.

22 HIDB, *Island of Mull: Survey and Proposals for Development*, 1973, 7.

23 HIDB, *Report for 1967*, 30.

24 HIDB, *Report for 1976*, 95–7. *Hansard*, Commons, 17 June 1965.

25 *WHFP*, 21 November 1975.

26 HIDB, *Report for 1970*, 5. Grassie, *Experiment*, 93–4.

27 Grassie, *Experiment*, 94.

28 A. Carty and J. Ferguson, 'Land' in A. Carty and A. MacCall Smith (eds), *Power and Manoeuvrability*, Edinburgh, 1978, 67.

29 HIDB, *Proposals for Changes in the Highlands and Islands Development Act*, 1978, 7–11. Grassie, *Experiment*, 95–6.

30 *WHFP*, 30 June 1978.

31 *Hansard*, Commons, 7 May 1980.

32 J. Ennew, *The Western Isles Today*, Cambridge, 1980, 50.

33 *Hansard*, Scottish Standing Committee, 16 April 1961.

34 *Stornoway Gazette*, 30 April 1963.

35 HIDB, *Report for 1977*, 9. D. R. F. Simpson (ed), *Island and Coastal Communities*, Glasgow, 1980, 53.

36 Simpson, *Communities*, 5.

37 CC, *Report for 1977*, 6. *WHFP*, 7 October 1977.

38 *WHFP*, 19 October 1979.

Chapter Seven. A Voice For Crofters

1 Proceedings of the assessors conference, 8 November 1979. As circulated by the Crofters Commission.

2 *WHFP*, 8 October 1976, 18 November 1977.

3 Angus Macleod in a letter to the author.

4 Federation of Crofters Unions to MacAskill, 16 November 1982: SCU Records.

5 *WHFP*, 4 April 1977. J. Hunter, *A Reformed Crofters Union*, HIDB, 1984, 30. Statistical information in what follows from this source.

6 Hunter, *Reformed Crofters Union*, 19.

7 Crofters Union Steering Group, *Bulletin No 1*, November 1984: SCU Records.

8 *The New Crofters Union*, 1985: SCU Records.

9 *Press and Journal*, 8 June 1985.

10 Crofters Union Steering Group, Submission to HIDB, April 1985: SCU Records.

11 *Crofter*, October 1985.

12 *Crofter*, February 1991.

13 Macleod to Hunter, 23 February 1985: SCU Records.

14 *Crofter*, August 1986.

15 Ibid.
16 *Crofter Housing: The Way Forward*, SCU, 1987.
17 *Crofter*, September 1989.
18 *Crofter*, June 1988.
19 *Crofter*, September 1987.
20 Sanderson speaking at SCU conference, Liniclate, 16 March 1989.
21 Ibid.

Chapter Eight. The Case For Crofting
 1 Congested Districts Board, *First Report*, 1899, Appendix III.
 2 Gillanders, 'Economic Life', 96–7.
 3 Leneman, *Fit for Heroes*, 204.
 4 D. J. MacCuish, 'Ninety Years of Crofting Legislation', *Transactions of the Gaelic Society of Inverness*, 1978, 230.
 5 Gillanders, 'Economic Life', 97, 104–5, 111.
 6 Bulletin of the European Communities, Supplement 4/88, *The Future of Rural Society*, Luxembourg, 1988, 8.
 7 *Future of Rural Society*, 21.
 8 J. Hunter, 'Against the Grain' in K. Cargill (ed), *Scotland 2000*, BBC Scotland, 1987, 93.
 9 *Highlander*, 29 August 1879.
10 *Crofter*, June 1989.
11 *Crofter*, June 1987.
12 *Crofter*, April 1986.
13 J. M. Bryden, 'Crofting in the European Context', *Scottish Geographical Magazine*, September 1987, 103.
14 *Crofter*, June 1987.
15 RSPB, *The Reform of the Common Agricultural Policy*, Sandy, 1988, 55.
16 *Hansard*, Commons, 27 January 1955. Highland Panel briefing paper, June 1948: SRO, SEP12/118. TC, Evidence: SRO, AF81/11.
17 TC, *Report*, 9.
18 *Crofter*, June 1988.
19 Ibid.
20 *Highlander*, 31 July 1875.
21 *Scottish Highlander*, 30 September 1886.
22 *Highlander*, 24 January 1874, 18 December 1875, 29 July 1876, 27 October 1877, 27 June 1879.
23 Darling, *Island Farm*, 199.
24 Geddes, *Lewis and Harris*, 276.
25 TC, Evidence: SRO, AF81/7.
26 TC, Evidence: SRO, AF81/23.
27 TC, *Report*, 9.

INDEX